INCLUDING LONDON UNDERGROUND

SECOND EDITION

Robert Pritchard & Alan Yearsley

Published by Platform 5 Publishing Ltd,
52 Broadfield Road, Sheffield, S8 0XJ, England.

Printed in England by The Lavenham Press Ltd, Lavenham, Suffolk.

ISBN 978 1 909431 69 0

▲ One of the most scenic routes to photograph trams is Manchester Metrolink's Oldham loop. On 19 April 2019 3004 approaches Newhey with a service from East Didsbury to Rochdale Town Centre. Metrolink operates 120 of these Bombardier M5000 trams, with a further 27 on order for delivery 2020–21. The first M5000 was delivered in 2009. **Robert Pritchard**

CONTENTS

Front Cover Photograph: Northern Line 1995 Stock 51721/51637 arrives at Kennington with a terminating service on 5 April 2019. **Robert Pritchard**

Back Cover Photograph (top): South Yorkshire Supertram tram-train 399 201 arrives at Meadowhall South/Tinsley with a service for Cathedral on 4 March 2019. **Robert Pritchard**

Back Cover Photograph (bottom): The first new Stadler train for the Glasgow Subway, No. 301, on display at the InnoTrans trade show in Berlin on 18 September 2018. **Robert Pritchard**

FOREWORD TO THE SECOND EDITION

Welcome to the second edition of the Platform 5 guide to the rolling stock of London Underground and Britain's light rail systems. Thanks to readers for the positive feedback received on the first edition of this book, something Platform 5 had been planning to do for a number of years. For a long time basic fleet details of Britain's light railway systems had been listed in the annual Platform 5 publication "British Railways Locomotives & Coaching Stock". These listings were never entirely satisfactory in terms of the level of detail provided, nor did they fall within the specific remit of that publication, hence the appearance of this book dedicated to London Underground and trams. As well as passenger carrying vehicles, details of the engineers' fleets and on-track machines are included.

Our aim with this book is to provide a comprehensive guide to the fleets, including not just fleet lists but also an overview of operations on each system and other useful information including brief history, route details, service frequencies and fare details for each system. This edition has been fully updated in include the latest information. Since the first edition we have seen the extension of the South Yorkshire Supertram network to Rotherham using tram-train vehicles. The first of the new Glasgow Subway trains has been delivered whilst a further 27 new trams have been ordered for Manchester Metrolink. CAF has been selected to supply 43 new full-length trains to Docklands Light Railway to replace most of its existing fleet of trains which date from 1991–2003 and new vehicles will also soon be ordered for Tyne & Wear Metro and West Midlands Metro. Route extensions are still happening as well – particularly in the West Midlands and in Manchester. And a decision was made shortly before this book went to press to extend the Edinburgh tram network to Newhaven, along its originally planned route.

New for this edition is the inclusion of hauled coaching stock as part of an expanded section on Preserved Underground Stock.

As ever we have both enjoyed travelling around sourcing new photos for this book, and we would also like to thank all of the photographers who have assisted with supplying photos for use in this book. If readers have any photos they would like to be considered for the next edition please send these to the email shown below: we are in particular looking for good quality London Underground photos including photos of the non-passenger (engineers and departmental) stock, as well as preserved stock. We are always of the opinion that the inside of a train (or a tram) is as important as the outside, so are pleased to also be able to include a number of interior photos in this edition.

ACKNOWLEDGEMENTS & CONTACT DETAILS

Thanks are given to all the individuals and organisations who have helped in the compilation of this book. Special thanks go to Peter Hall, particularly for his work in tracking down preserved stock and guiding us through the area of On-Track Machines. Also thanks to Ian Beardsley for his research on the maps and to Trackmaps for its cartography.

The authors would welcome updates and corrections to this book, of which readers have first-hand knowledge. Please send any notifications to Robert Pritchard at the Platform 5 address on the title page, or by email to updates@platform5.com (Tel 0114 255 2625).

This book is updated to information received by July 2019.

UPDATES

Updates to Platform 5 UK handbooks can be found in the monthly magazine **Today's Railways UK**, which is the only UK railway magazine to publish official Platform 5 stock changes as well as a separate column dedicated to the latest Light Rail News. **Today's Railways UK** is available from all good newsagents or on post-free direct subscription. Please see the inside cover of this book for further details.

INTRODUCTION

This book lists the passenger-carrying vehicles of all ten public-carrier light rail and metro systems in the UK. Locomotives and on-track machines that fall within the custodianship of each system and are permitted to operate under their own power or in train formations throughout those systems are also included.

Locomotives, road-rail vehicles and other maintenance machines normally only permitted to be used within depot confines or within engineering possessions are not included, nor are contractors' locomotives and on-track machines that may see use on the networks for short periods.

Vehicles that previously saw service on the currently active networks but are now considered to be preserved are included. Historic tramway vehicles that still exist but spent their working lives on now-closed tramways in the UK are not included.

GENERAL NOTES

This book has been divided into three main sections covering The London Underground, Preserved Underground Stock and UK Light Rail and Metro Systems. Within the London Underground section, each line has its own sub-section; these are listed broadly in chronological order of age of rolling stock used on each line, with tube lines preceding Sub-Surface lines.

For every LU line in section 1 and light rail undertaking in section 3, a general introduction provides a brief history of each line and details of day-to-day operations including service frequencies, operating times and depot/maintenance information. Further details of fares and ticketing requirements are also provided and where appropriate future plans for each line or network.

This introduction is followed by detailed fleet lists with tabulated technical data and unit formations where appropriate. Detailed technical specifications should be read with reference to the following notes and also the appendices at the back of the book.

NUMBERS AND NAMES

There is no universal numbering scheme in place for UK metros and light rail systems, each system instead adopting their own numbering scheme. Vehicles are listed within each sub-section in chronological order of current running number. Notes regarding individual numbering schemes are provided within each sub-section as appropriate. Former numbers carried by preserved vehicles are shown. The current name of vehicles, where carried, is shown.

LOCATIONS

Details of the depots used for routine servicing and maintenance are provided in the introduction to each sub-section; depot allocations are not in use. For preserved items the location at which the item can usually be found is given, although it should be noted that some vehicles, especially locomotives, sometimes visit other locations for operation, display or mechanical attention. A list of locations with OS grid references can be found in Appendix I: List of Locations.

WHEEL ARRANGEMENT

For modern metro and light rail vehicles (and main line diesel and electric locomotives) the system whereby the number of powered wheels on a bogie is denoted by a letter (A = 1, B = 2, C = 3 etc) and the number of unpowered axles is denoted by a number is used. The letter "o" after a letter indicates that each axle is individually powered and a + symbol indicates that the bogies are intercoupled.

The Whyte notation is used for steam locomotives and diesel shunting locomotives with coupled driving wheels. The number of leading wheels are given, followed by the number of driving wheels, then the number of trailing wheels. Suffixes are used to denote tank locomotives as follows: T = side tank, PT = pannier tank, ST = saddle tank, WT = well tank. For example 2-6-2T. 4w indicates a four-wheeled non-bogied vehicle.

DIMENSIONS

The units of measure used in this book are those most appropriate to current usage. In most cases this is the metric system except where the imperial system is still in common use. For example, length, height and width dimensions are usually given in metric metres, but maximum speed is given in imperial miles per hour. In most instances weights are stated in metric tonnes, except in the case of a few very old vehicles where imperial tons are used.

Steam locomotive cylinder dimension are given in imperial feet and inches; the diameter is given first, followed by the stroke. (I) indicates two inside cylinders, (O) two outside cylinders.

ACCOMMODATION

The number of seats in each vehicle is given. First Class accommodation is not provided in any vehicles listed in this book (except for some of the preserved vehicles) so all seats can be regarded as a single classification.

BUILD DETAILS

For each vehicle the builder and year of construction are given. A full list of builders can be found in Appendix IV: Private Manufacturers.

▲ A train of Bakerloo Line 1972 Stock, the oldest still in passenger service on the Underground network, approaches Willesden Junction with a service for Elephant & Castle on 2 September 2017. Car 3256 is leading. **Keith Fender**

1. LONDON UNDERGROUND

INTRODUCTION

The London Underground is the world's oldest metro system covering a total of 11 lines, with a history going back to 1863 when the Metropolitan Railway opened its first stretch of line between Paddington and Farringdon. Much of the rest of the sub-surface network also dates from between the 1860s and the early 20th century, and the first deep level tube line opened in 1890 between Stockwell and King William Street on what is now the Northern Line. The largest metro system in Europe, today London Underground serves a total of 270 stations and covers 402 km (250 miles) of track.

In total there are 11 Underground lines, each shown in their own colour on the official map. Four lines, the Circle, District, Hammersmith & City and Metropolitan are Sub-Surface Lines (SSLs), since they operate above ground or just below ground and use trains built to the same loading gauge as rolling stock on the National Rail network (with which they share tracks in a number of locations). The other seven Underground lines (Bakerloo, Piccadilly, Central, Waterloo & City, Northern, Jubilee and Victoria) are deep level tube lines running in tunnels bored rather than built by the "cut and cover" method used for sub-surface tunnels. The "cut and cover" method caused a great deal of surface disruption during construction so towards the end of the 19th century a new system of tunnelling was developed that involved boring a circular tunnel at deep level and lining it with cast iron segments. Lines built in this way have a separate tunnel for each direction of running and because of the small diameter it is only possible to accommodate trains of very restricted dimension – the Tube stock. Most of the deep level tube lines do also include some above ground sections in the suburbs, however, although the entire LU network is often colloquially referred to as "The Tube". Despite its name, only 45% of the network is in tunnels. This section contains separate more detailed histories and descriptions of each line.

London's underground railway system developed rather haphazardly, creating traffic as it grew without any central control or planning, except that nominally exercised by Parliament, for almost 70 years until the formation of the London Passenger Transport Board (LPTB) in 1933 (although the brand name Underground first started to be displayed at station entrances in 1908). As a result of this the various lines appeared to have little in common apart from the standard track gauge of 1435 mm: there were different tunnel sizes, different platform lengths, different types of signalling and even different current rail systems. All these variations gave rise to different types of rolling stock. Within the limits of cost and time, considerable standardisation has been achieved, as the new S Stock on the Sub-Surface lines and 1995/1996 Tube stocks on the Northern and Jubilee lines show. The forthcoming new Siemens "Inspiro" trains for the deep tube lines will allow further standardisation.

Subsequent political changes saw LPTB evolve through the London Transport Executive (LTE) in 1948, London Transport Board (LTB) in 1963, London Regional Transport (LRT) in 1984, and now London Underground Ltd (LU), part of Transport for London (TfL) which was formed in 2000 when the Greater London Authority was established along with the directly elected Mayor of London who is responsible for all forms of public transport (apart from most National Rail services). The terms London Transport and London Underground were commonly used for all these organisations, and LT and LU are used interchangeably within this book.

An extension to the Northern Line is currently being constructed, a 2 mile branch from Kennington to Battersea, which will all be in tunnel. With one intermediate station at Nine Elms this extension is due to open in September 2021 and will be served by the existing Northern Line fleet.

THE TRAIN FLEET

London Underground operates seven basic train types, each line generally operating a different type of train (the exceptions being 1992 Stock of a similar type used on both the Central and Waterloo & City lines and the S Stock universally used on all the Sub-Surface Lines).

Official LU rolling stock designation is by the year in which manufacture started for Tube stock, and by arbitrary letter with year suffix for Sub-Surface stock, e.g. the Central Line's 1992 tube stock is abbreviated to 92TS. Naturally there are exceptions to every rule – 1995 and 1996 Tube Stock entered service the opposite way around, while S Stock is referred to as S7 or S8, denoting the number of cars in the train, not the year of manufacture.

Tube stock is built to a much lower overall height than Sub-Surface stock to fit the more restricted loading gauge of the deep level tunnels. The height of Tube stock varies between 2.869 m for 1992 Stock used on the Central and Waterloo & City Lines and 2.888 m for Piccadilly Line 1973 Stock (see individual entries for details) to fit the differing sizes of tunnels built by different railway companies. The now withdrawn D78 Stock had a height of 3.630 m, and the height of S7 and S8 Stock vehicles is 3.682 m.

Despite the long period of central control and deployment of large successive standardised fleets such as 1920s "Standard" stock, 1938 Stock, and 1959/1962 Stock common to several lines, today's situation is largely reversed with dedicated captive fleets on each line. The notable exception is the S Stock now common to all four Sub-Surface lines (although 8-car sets are limited to the Metropolitan Line with 7-cars used on the other three SSLs).

In this book rolling stock is listed chronologically by the official designation of all types. Therefore the order is 1972, 1973, 1992, 1995, 1996, 2009 and then S8/S7.

In January 2016 TfL issued an Invitation to Tender for a new Deep Tube train that will eventually be rolled out across the Piccadilly, Central, Bakerloo and Waterloo & City Lines over the next 10–15 years. In June 2018 Siemens was awarded an initial £1.5 billion contract to build 94 new "Inspiro" trains for the Piccadilly Line, with options for further trains. It is envisaged that around 250 trains will be needed for all four of the lines that will see the new trains (around 100 for the Central Line, 40 for the Bakerloo Line and 10 for the Waterloo & City Line). The Piccadilly Line trains will be built both at Siemens' factory in Vienna and at a new factory under construction in Goole, UK. The first trains are due to be delivered for testing in 2023, with service introduction between 2024 and 2026. The "Inspiro" trains will be the first on the deep level tube lines to feature air conditioning and walk-through gangways. The full technical specification for the trains has not yet been made public.

Most present stock types have two thirds-to-three quarters of axles motored within a full train, but 1992 and S Stocks have all axles motored. 1972, 1973 and 1992 Stock have traditional series-wound Direct Current (DC) commutator traction motors and the newer 1995, 1996, 2009 and S Stocks, three-phase Alternating Current (AC) synchronous motors. 1972 and 1973 traction motors are nose-suspended while 1992, 1995, 1996, 2009 and S Stocks have bogie-mounted motors and flexible drives via a gearbox. 1972 and 1973 Stocks have conventional camshaft resistance controls, 1992 chopper, 1996 GTO, and 1995, 2009 and S Stock are IGBT.

Again, despite much standardisation over many years, successive developments in train protection systems have also influenced the trend towards line specific captive stock fleets. All LU trains are equipped with automatic train protection and have been for many years. Train borne tripcocks and track side train stops form the traditional system and are still in use on the Bakerloo, Piccadilly and all Sub-Surface Lines. The then new Victoria Line opened in 1967 with a unique LT developed system. Central, Jubilee and Northern Lines followed in the 21st Century and the original Victoria Line system has now been replaced. Each line specific train protection system is listed.

Unlike on the National Rail network, LU stock is not generally formed of fixed formation trains, the exception being the new 2009 and S Stock, plus the Northern Line 1995 Stock which is now generally used in fixed formations. Other units are generally formed of motor and trailer cars semi-permanently coupled. Units, even those in fixed formations, do not have set numbers, unlike on the national network, although they are sometimes referred to as "Train 1" etc.

All trains currently in service have bar couplers between cars and LU automatic Wedgelock between units, except S Stock which have flange muff couplers between cars. In addition S Stock auto couplers have pneumatic connections.

Automatic Train Operation (ATO), whereby the train effectively drives itself and the driver simply opens and closes the doors and presses a start button, is currently in use on the Central, Jubilee, Northern and Victoria Lines (the Victoria Line having used ATO since its opening). At the time of going to press ATO was being commissioned on the Circle and Hammersmith & City Lines, and was due to be extended to the District and Metropolitan Lines by 2022. The Bakerloo, Piccadilly and Waterloo & City Line trains continue to be driven manually, and this is expected to be the case until the introduction of "Inspiro" trains on these lines.

Maximum speed quoted is the line limit; the stock may well have a higher design speed.

All LUL passenger trains are electrically powered vehicles which are driven on truck-mounted motors. Not all cars have motors: these unmotored cars are termed "trailers". The following Type Codes are used throughout the book:

DM Driving Motor = powered car with driving cab;
NDM Non Driving Motor = powered car without driving controls; see also;
M Non Driving Motor
M1 Non Driving Motor 1st variant (S Stock);
M2 Non Driving Motor 2nd variant (S Stock);
MS Non-Driving Motor fitted with Sandite equipment (S Stock);
T Trailer = unpowered car without driving controls;
UNDM Uncoupling Non Driving Motor = powered car with shunting driving control.

De-icing cars are marked with a blue circle next to the number and these are listed in this publication. When actually de-icing a small blue external light is lit.

All passenger carrying LU trains carry a standard livery. This is principally white, with a wide blue lower bodyside band, a red front end and red doors.

In photo captions in the LU section of this book the full train formations are given where known, listed by the lowest fleet number of that set/half set/quarter set.

TRACTION SUPPLY

Traditional LU traction supply is a nominal 630 V DC from third and fourth rail – but upgrading to 750 V DC is taking place on some lines, dependent on the ability of newer trains to accept the higher voltage. All Sub Surface Lines (Metropolitan, District, Circle and Hammersmith & City) have now been upgraded to 750 V DC, except where electric traction supply sections are shared with non S Stock trains. It is a long-term aspiration to convert the whole network to the higher voltage.

The current is collected by the trains from two conductor rails laid on each track in addition to the running rails, known as a four-rail DC system. The conductor rails are insulated from earth by porcelain "pots" mounted on the sleepers. Where points or crossings occur in the running rails, gaps are left in the current rails to allow the train wheels to pass freely through the junction and to ensure the shoegear of trains using the conflicting route does not foul it.

MAINTENANCE

Almost all maintenance work on LU trains is carried out at depots, but individual components such as motors, motor alternators, compressors and wheelsets units may be sent away for major overhaul, either to the Railway Engineering Works (REW) at Acton or to outside contractors.

The transfer of heavy overhaul work to the maintenance depots has changed the type of work undertaken at Acton. Instead of carrying out heavy overhauls most of the work is now repairing and overhauling components sent by road from the depots.

It should be noted that the 1995 and 1996 Tube stocks have a different ownership and maintenance regime to the rest of the LU fleet. The 1995 Stock is owned by HSBC/RBS or Alstom and leased to LU. The trains manufacturers, Alstom, has a contract to maintain the trains at Morden and Golders Green. The 1996 Stock is owned by London Underground but a maintenance contract is in place with Alstom to maintain the trains at Stratford Market depot.

Finally, the depot at Ruislip is Transplant Rail Operations engineering train base depot.

PASSENGER NUMBERS

The 2017–18 annual line by line passenger figures for LUL are as follows:

Northern Line: 291.7 million
Central Line: 282.4 million
Jubilee Line: 271.4 million
Victoria Line: 265.4 million
District Line: 222.0 million
Piccadilly Line: 202.2 million
Circle and Hammersmith & City: 134.7 million
Bakerloo Line: 113.9 million
Metropolitan Line: 83.8 million
Waterloo & City Line: 13.6 million

Total: 1357 million (2017–18). In 2015–16 the total was 1349.3 million.
(There may be discrepancies between the line by line figures and the total owing to rounding.)

TICKETING AND FARES

The LU and National Rail networks within Greater London are divided into nine fare zones, and fares are calculated according to the number of zones through which the passenger travels. When making just one or two single journeys on the Underground in a day, the best value fares can be obtained by tapping a TfL "Oyster card" smartcard or a contactless credit or debit card on the ticket gate or card reader at the station entrance before boarding and at the exit after alighting. Ordinary paper tickets are still available for a single journey, but are much more expensive than when using an Oyster card or contactless payment card. In 2019 a single journey within Zone 1 costs £2.40 with an Oyster or contactless card, or £4.90 with a paper ticket.

In recent years most ticket offices have been closed at LU stations. Oyster cards can be obtained from ticket machines at stations, TfL Visitor Centres at key stations in central London, newsagents and convenience stores displaying the "Oyster Ticket Stop" sign, or ordered online via the TfL website. Many passengers increasingly may prefer to use a contactless bank card, as

▲ In November 2018 Siemens won the contract to build the next generation of Deep Tube trains, starting with 94 sets for the Piccadilly Line. This is an artist's impression of what the new Piccadilly Stock will look like. **Courtesy Siemens**

fare payments will still be processed in exactly the same way as with an Oyster card provided that the same card is used to touch in and touch out on the same journey. Oyster cards can be topped up at ticket machines at stations and at Oyster Ticket Stop shops. They will only work if the card holder has enough credit on the card for the journey being made.

Visitors to London can buy a Visitor Oyster card in advance before travelling to London. These cost £5 and come pre-loaded with an amount of credit of the user's choice between £10 and £50. They work in the same way as an ordinary Oyster card.

Paper One Day Travelcards are also still available, and are recommended for exploring London, especially if you plan to spend any length of time on stations observing trains, as journeys using Oyster or contactless cards are subject to a time limit, and the maximum fare will be charged if you exceed this limit or forget to touch out at the end of the journey. However, unfortunately paper Zones 1–2 and off-peak Zones 1–4 Travelcards have now been abolished, so only the Zones 1–6 version is still offered (plus an Anytime version for Zones 1–4). In 2019 the Zones 1–6 Travelcard costs £13.10 off-peak (or Zones 1–4 Anytime) and the Anytime Zones 1–6 version costs £18.60. Railcard holders (weekends and bank holidays only in the case of Network Railcards) can buy an off-peak Zones 1–6 Travelcard for £8.60. There is also a version available for Zones 1–9 at £23.50 Anytime or £13.90 off-peak (£9.20 for Railcard holders).

If you do choose to use an Oyster or contactless card for several journeys on the same day, you will receive an automatic price cap once you have used a certain amount of credit; in most cases this is the same as, or slightly less than, the cost of an equivalent One Day Travelcard for the journeys that you have made. Zones 1–2 and 1–4 price caps do still apply despite the withdrawal of paper Travelcards covering only these zones.

A paper Travelcard bought on a Friday is valid on Night Tube services until 04.30 on the Saturday, and similarly one purchased on a Saturday will be valid on Night Tube services until 04.30 on the Sunday.

Paper Travelcards, Oyster and contactless cards are also valid on almost all other forms of public transport in Greater London: buses, National Rail services, London Overground, the Docklands Light Railway, and London Tramlink (Oyster and contactless cards can also be used on River Bus boats and the Emirates Air Line cable cars).

Greater London residents are entitled to a 60+ Oyster Photocard when aged 60–65 or a Freedom Pass when aged 66 or over (or disabled). Both of these passes allow free travel on the Underground, buses, DLR, Tramlink and London Overground at all times (Freedom Passes are also valid on National Rail services outside the morning peak period). Pensioners and disabled people from other parts of England (but not from Scotland, Wales or Northern Ireland) may use their concessionary travel passes for free travel on London buses, but NOT on the Underground or any other modes of transport in London.

Full details of fares and tickets available can be found on the TfL website (www.tfl.gov.uk) and on the independent Oyster Rail site: www.oyster-rail.org.uk.

PHOTOGRAPHY

Once again, the authors have enjoyed gathering recent photographs of the London Underground to illustrate this book, and with the odd exception, have been able to photograph Underground trains with no problems. Photography and filming is permitted on LU stations and trains provided that you are doing so only for non-commercial purposes. However, for safety reasons flash photography and the use of tripods are prohibited under Section 4.5 of the TfL Conditions of Carriage (see http://content.tfl.gov.uk/tfl-conditions-of-carriage.pdf). Photography and filming permits are available from the LU Film Unit at £50 per month for commercial photographers.

The best advice is to be sensible. Talk to station staff if they approach you and explain what you are doing. It may seem obvious, but good photos of Underground trains are not going to be obtained at central London stations in the rush hour, so choose locations and time carefully. There have been occasional reports of station staff wrongly claiming that all photography and filming is prohibited. If this happens, you may like to politely point out that the Conditions of Carriage only ban the use of flash and tripods, and if necessary ask the member of staff involved to speak to their supervisor to clarify the position. You may wish to approach the on-duty Station Supervisor wherever possible and check that it is OK to take photos. If permission is refused, there may be a good reason for this, such as at busy times or if an event is taking place on or near the station which is deemed to require tightened security measures.

▲ Gants Hill on the Central Line was a 1940s station designed by Charles Holden in the style of one of the Moscow Metro stations, with a high vaulted roof. On 3 April 2019 91343 brings up the rear of a southbound service. **Robert Pritchard**

▼ Victoria Line 2009 Stock at Walthamstow depot on 16 February 2019. **Keith Fender**

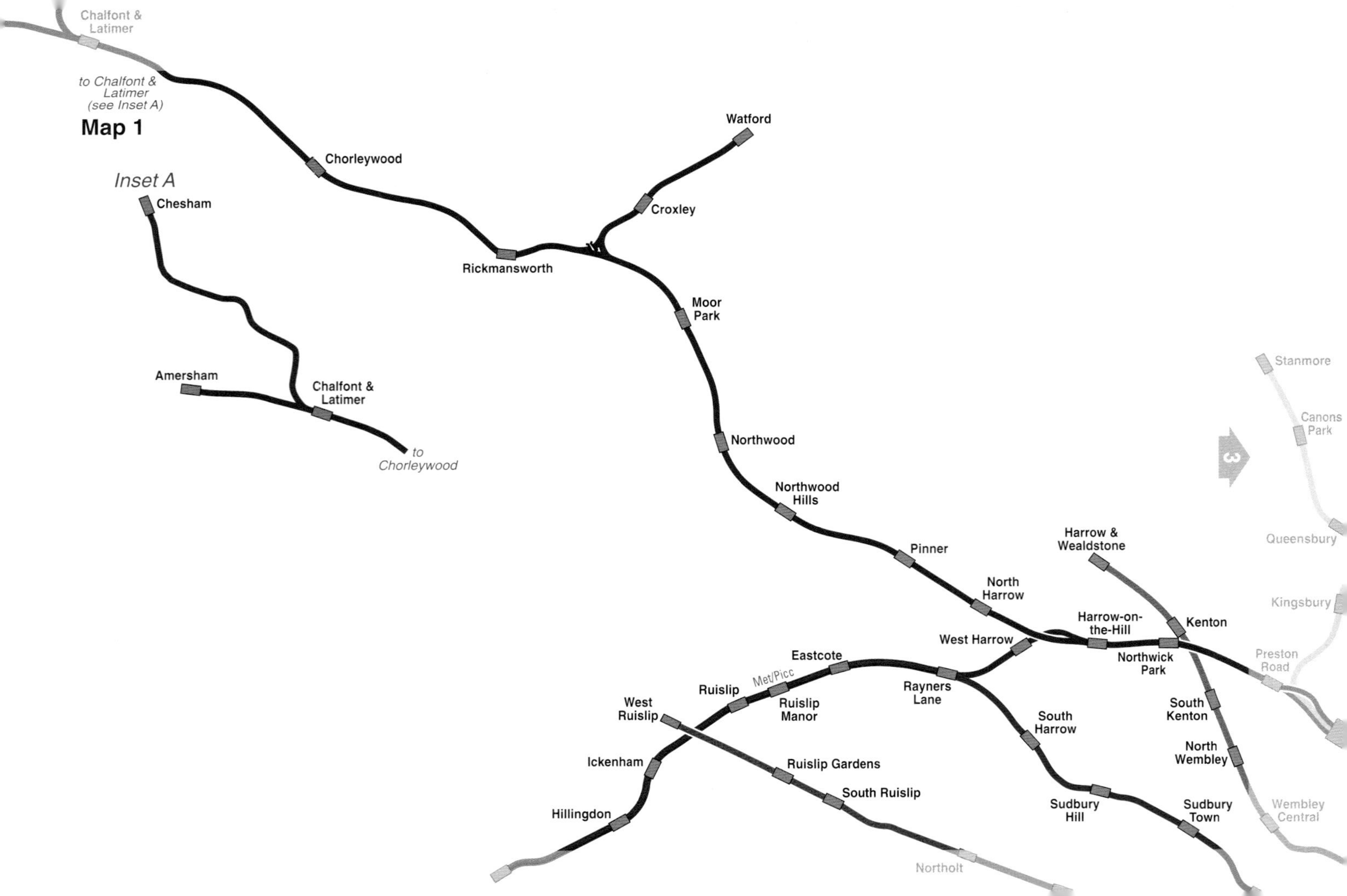

Map 1
Chalfont & Latimer
to Chalfont & Latimer (see Inset A)
Chorleywood
Watford
Croxley
Rickmansworth
Moor Park
Northwood
Northwood Hills
Pinner
North Harrow
Harrow & Wealdstone
Harrow-on-the-Hill
Kenton
Northwick Park
West Harrow
Eastcote
Rayners Lane
Ruislip
Met/Picc
Ruislip Manor
West Ruislip
Ickenham
Hillingdon
Ruislip Gardens
South Ruislip
Northolt
South Harrow
Sudbury Hill
Sudbury Town
South Kenton
North Wembley
Wembley Central
Preston Road
Kingsbury
Queensbury
Canons Park
Stanmore
3
Inset A
Chesham
Amersham
Chalfont & Latimer
to Chorleywood

Map 2

LONDON UNDERGROUND

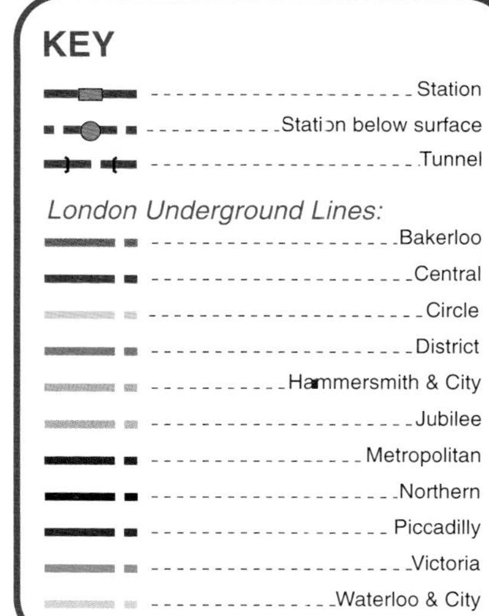

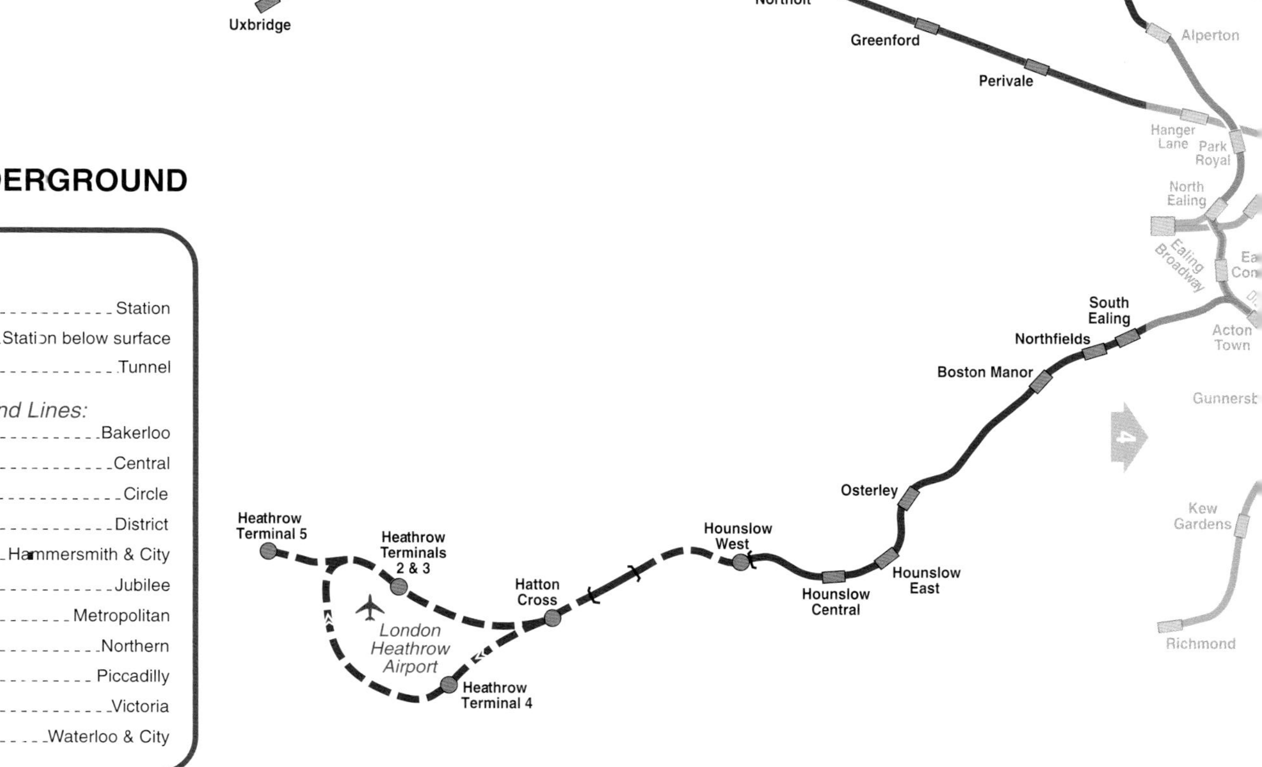

Map 3

High Barnet
Totteridge & Whetstone
Woodside Park
West Finchley
Mill Hill East
Finchley Central
East Finchley
Highgate
Archway
Tufnell Park
Kentish Town
Cockfosters
Oakwood
Southgate
Arnos Grove
Bounds Green
Wood Green
Turnpike Lane
Manor House
Finsbury Park
Arsenal
Holloway Road
Caledonian Road
Highbury & Islington
Seven Sisters
Tottenham Hale
Blackhorse Road
Walthamstow Central
Leyton
Stanmore
Canons Park
Queensbury
Kingsbury
Edgware
Burnt Oak
Colindale
Hendon Central
Brent Cross
Golders Green
Hampstead
Belsize Park
Chalk
Swiss Cottage
Kenton
Preston Road
Wembley Park
Met/Jub
Neasden
Dollis Hill
Willesden Green
Kilburn
West Hampstead
Finchley Road
North
Sudbury Town
Wembley Central
Stonebridge Park
1
5

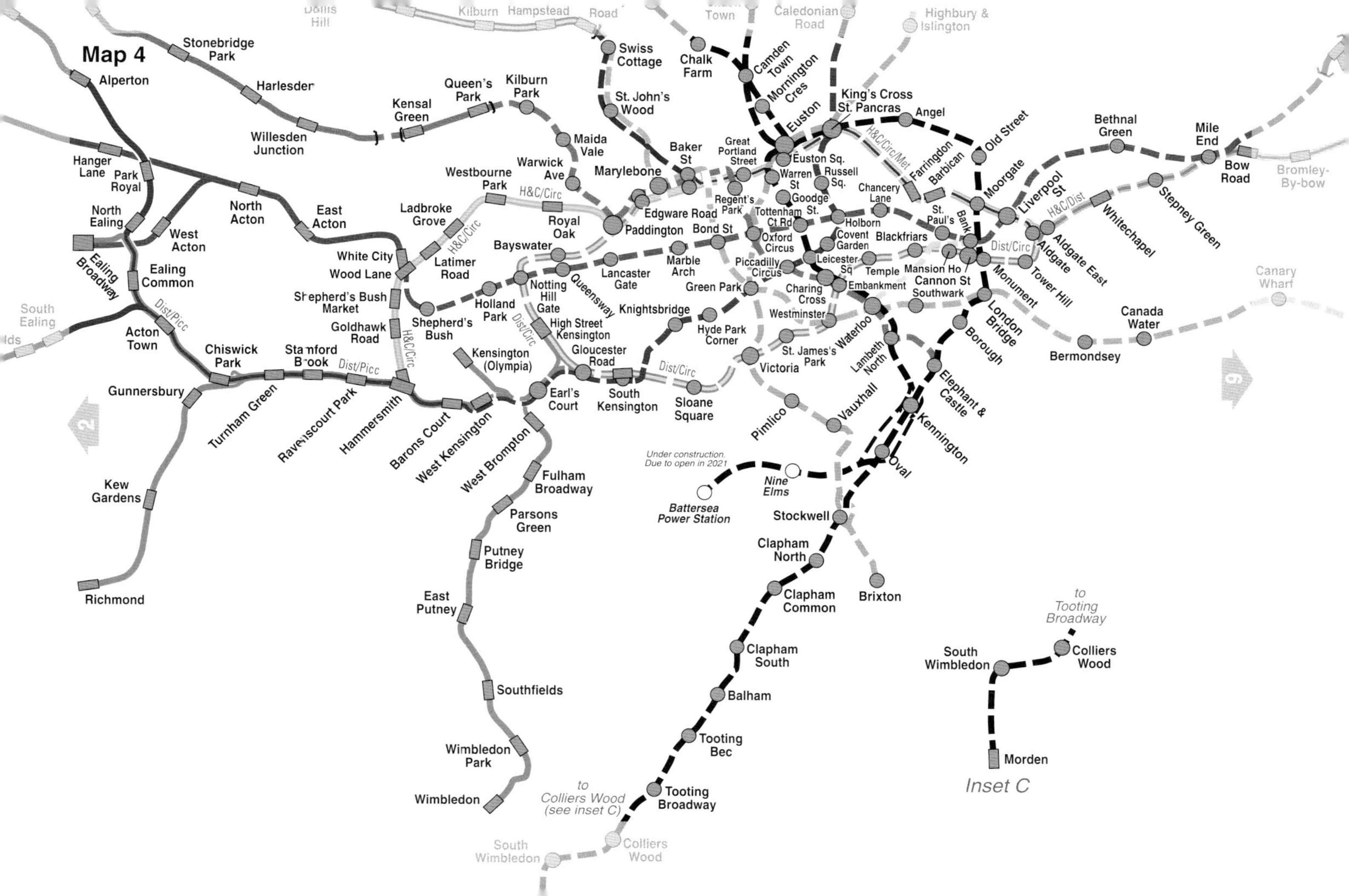

Map 4
Alperton
Stonebridge Park
Harlesden
Willesden Junction
Kensal Green
Queen's Park
Kilburn Park
Maida Vale
Warwick Ave
Westbourne Park
Swiss Cottage
St. John's Wood
Chalk Farm
Camden Town
Mornington Cres
Euston
King's Cross St. Pancras
Angel
Old Street
Bethnal Green
Mile End
Bow Road
Bromley-By-bow
Hanger Lane
Park Royal
North Ealing
Ealing Broadway
West Acton
North Acton
East Acton
Ealing Common
Acton Town
South Ealing
Chiswick Park
Gunnersbury
Kew Gardens
Richmond
Turnham Green
Ravenscourt Park
Stamford Brook
Goldhawk Road
Shepherd's Bush Market
Wood Lane
White City
Latimer Road
Ladbroke Grove
Shepherd's Bush
Holland Park
Notting Hill Gate
Queensway
Bayswater
Royal Oak
Paddington
Edgware Road
Marylebone
Baker St
Great Portland Street
Regent's Park
Warren St
Euston Sq.
Russell Sq.
Goodge St.
Tottenham Ct.Rd
Oxford Circus
Bond St
Marble Arch
Lancaster Gate
Knightsbridge
Hyde Park Corner
Green Park
Piccadilly Circus
Leicester Sq
Covent Garden
Holborn
Chancery Lane
Farringdon
Barbican
Moorgate
Liverpool St
Aldgate
Aldgate East
Whitechapel
Stepney Green
St. Paul's
Bank
Blackfriars
Mansion Ho
Cannon St
Monument
Tower Hill
Temple
Embankment
Charing Cross
Westminster
Southwark
Waterloo
London Bridge
Borough
Bermondsey
Canada Water
Canary Wharf
Highbury & Islington
Caledonian Road
Hammersmith
Barons Court
West Kensington
Kensington (Olympia)
High Street Kensington
Earl's Court
Gloucester Road
South Kensington
Sloane Square
Victoria
St. James's Park
Pimlico
Vauxhall
Lambeth North
Elephant & Castle
Kennington
Oval
Under construction. Due to open in 2021
Nine Elms
Battersea Power Station
Stockwell
Brixton
Clapham North
Clapham Common
Clapham South
Balham
Tooting Bec
Tooting Broadway
to Colliers Wood (see inset C)
Colliers Wood
South Wimbledon
West Brompton
Fulham Broadway
Parsons Green
Putney Bridge
East Putney
Southfields
Wimbledon Park
Wimbledon
H&C/Circ
H&C/Circ/Met
H&C/Dist
Dist/Circ
Dist/Picc
2
6
to Tooting Broadway
Colliers Wood
South Wimbledon
Morden
Inset C

Map 5

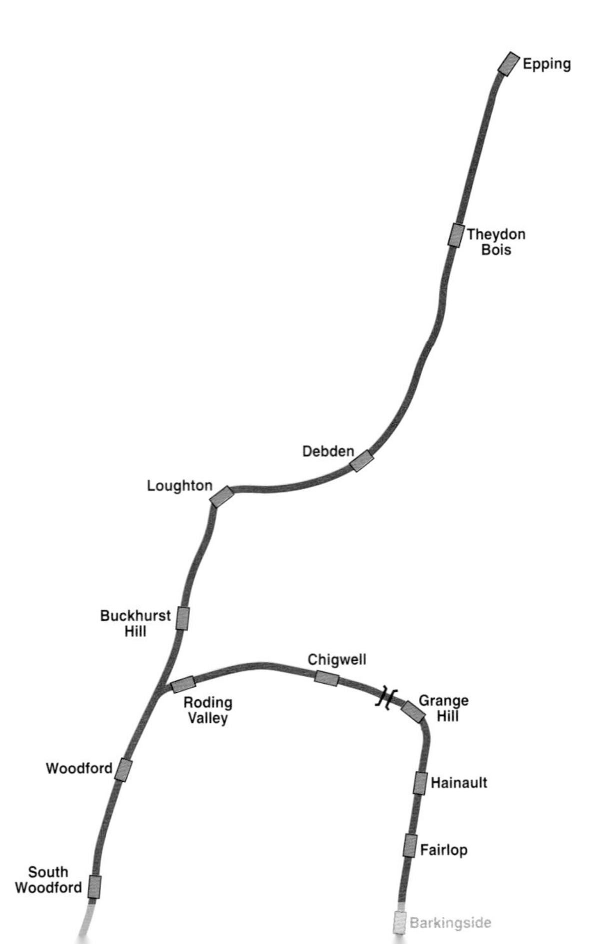

Epping
Theydon Bois
Debden
Loughton
Buckhurst Hill
Chigwell
Grange Hill
Roding Valley
Woodford
Hainault
Fairlop
South Woodford
Barkingside
3

Map 6

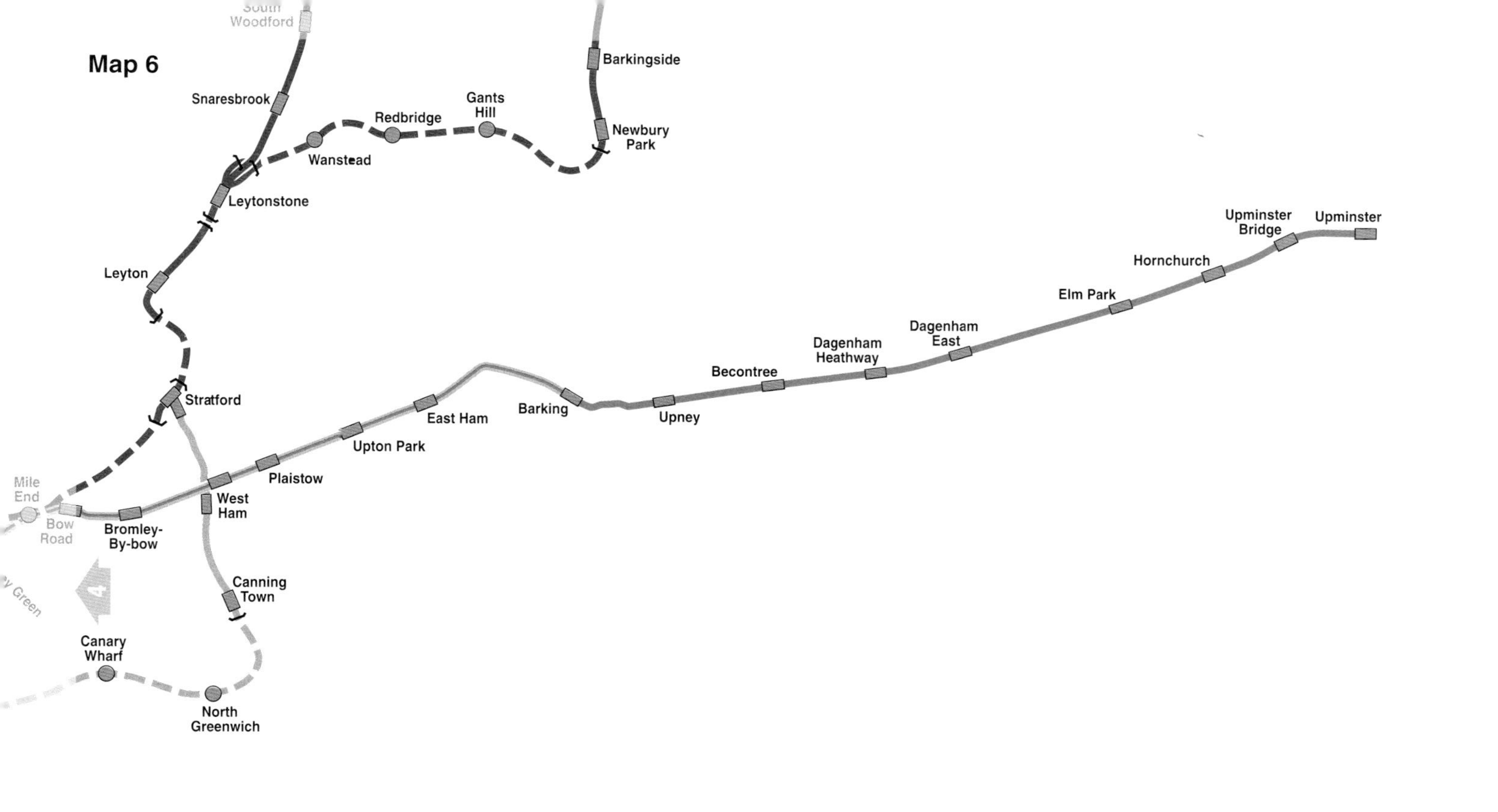

Woodford
Barkingside
Snaresbrook
Redbridge
Gants Hill
Newbury Park
Wanstead
Leytonstone
Upminster Bridge
Upminster
Leyton
Hornchurch
Elm Park
Dagenham East
Dagenham Heathway
Becontree
Stratford
East Ham
Barking
Upney
Upton Park
Plaistow
Mile End
West Ham
Bow Road
Bromley-By-bow
Canning Town
4
Canary Wharf
North Greenwich

1.1. BAKERLOO LINE

Route: Harrow & Wealdstone–Elephant & Castle. **Route length:** 14½ miles.
No. of trains: 36 x 7-car (252 cars). **No. of stations:** 25. **First section opened:** 1906.

The Bakerloo Line is one of the shorter London Underground lines, but provides an important link from north-west London to the centre and shopping district of Oxford Street, also linking the main line stations of Paddington, Marylebone (only served by the Bakerloo Line), Charing Cross and Waterloo. The southern terminus at Elephant & Castle is in Zone 1, and (apart from the Waterloo & City line) is the nearest LU terminus to the centre of London.

The Bakerloo Line has its origins in the Baker Street & Waterloo Railway from which it gets its name. Despite this the first section of the line, opened in 1906, actually ran from Baker Street to Elephant & Castle, two stations south of Waterloo, which has been the line's southern terminus since its opening. Construction of the line had started in 1898. The line was extended westwards from Baker Street to Edgware Road via Marylebone in 1907, then on to Paddington in 1913, Queen's Park and Willesden Junction in 1915, and Watford Junction in 1917, running alongside the West Coast Main Line (then owned by the London & North Western Railway) and sharing tracks with the LNWR's Watford DC Line north of Queen's Park.

From 1939 until 1979 the Bakerloo Line also included the Stanmore branch. Since 1979 this has been operated as part of the Jubilee Line (see separate section).

From 1965 Bakerloo Line trains only ran north of Queen's Park in peak hours, since this section was also served by Euston–Watford local trains on the DC Lines. In 1982 Bakerloo Line services were withdrawn north of Stonebridge Park, but were extended back as far as Harrow & Wealdstone in 1984 and restored to run an all day seven days a week service north of Queen's Park in 1989.

The Bakerloo Line south of Queen's Park to Elephant & Castle is all in tunnel, with the northern section from Queen's Park being above ground.

A number of schemes to extend the line have been considered over the years, but have so far never come to fruition. These include a northern extension from Edgware Road to Cricklewood mooted in 1908, and southern extensions to Camberwell proposed in 1931 and 1947. Network Rail's 2011 London & South-East Route Utilisation Strategy envisaged an extension from Elephant & Castle to Lewisham and then on to Hayes via Southeastern's existing line with a possible branch to Beckenham Junction and on to Bromley. A public consultation in 2015 showed strong support for the extension as far as Lewisham, for which Transport for London has now allocated [illegible]ds in its business plan with completion expected by 2028–29. Any extension beyond Lewisham would be considered as a separate phase at a later date.

SERVICES

The basic all-day service consists of 20 trains per hour on the core section between Elephant & Castle and Queen's Park, giving a train approximately every 3 minutes. Of these, six trains per hour run to and from Harrow & Wealdstone and a further three continue as far as Stonebridge Park. Unlike all other deep level tube lines (apart from the Waterloo & City) there is currently no Night Tube service on the Bakerloo Line. Trains run from around 05.30 to 00.30 on Mondays–Saturdays and from 07.00 to midnight on Sundays.

ROLLING STOCK AND DEPOTS

All Bakerloo Line trains are currently formed of 1972 Tube Stock, which are similar to the (now withdrawn) 1967 Stock used on the Victoria Line until 2011. 63 of these trains were originally built by Metro-Cammell, Birmingham for the Northern Line in two batches; Mark 1 and Mark 2. They were cascaded to the Bakerloo Line from 1977, but were concentrated on the Jubilee Line from 1979. Today 36 trains remain, mainly Mark 2 stock refurbished 1991–95 by Tickford, Rosyth. These trains have been in sole charge on the Bakerloo Line since the mid-1980s following the withdrawal of the earlier 1938 and 1959 Stock, and are expected to remain in service until the introduction of new fully automatic trains by 2033 as part of TfL's New Tube for London programme, by which time the existing fleet will be some 60 years old. Having been refurbished in the mid-1990s, 1972 Stock is currently receiving a life extension at Acton Works, which includes corrosion rectification, new flooring and car body repairs. This work is due for completion in 2020. As a separate project cars have also received an interior refurbishment involving the renewal of the lighting, handrails and seat moquette.

▲ Bakerloo Line 1972 Stock 3240/3547 runs alongside the West Coast Main Line, on the approach to South Kenton, with a service for Elephant & Castle on 9 December 2018. **Robert Pritchard**

▼ 3260/3550 arrives at Edgware Road with a service for Elephant & Castle on 16 March 2019. **Robert Pritchard**

The 1972 Stock operates as 7-car fixed formation trains, formed of one south end 4-car and one north end 3-car units, not in permanent formation. The 4-car unit south end DM cab at the centre of the formation is unused in traffic and is generally known as a MM (Middle Motor). 32 of the 36 trains are required for the peak service.

The main depot is at Stonebridge Park; trains also stable at Queen's Park, Elephant & Castle and London Road – near Lambeth North. A unique feature of the depot at Queen's Park is that passenger-carrying trains run through it. To reach Queen's Park depot from the north, or in case of disruption south of Queen's Park, some trains run onto the Euston–Watford DC Lines as far as Kilburn High Road, where the fourth rail remains energised, and then reverse.

1972 TUBE STOCK — BAKERLOO LINE

Formation: 7-car; DM–T–T–MM (* UNDM) + UNDM–T–DM.
Construction: Aluminium body on steel underframe.
Traction Motors: 4 x DC nose-suspended Brush LT115 of 53 kW (71 hp) per motor car.
Wheel Arrangement: Bo-Bo + 2-2 + 2-2 + Bo-Bo (+) Bo-Bo + 2-2 + Bo-Bo.
Braking: Electro-pneumatic and rheostatic.
Dimensions: 16.09/15.98 x 2.64 m. **Height:** 2.877 m.
Bogies: Plate frame.
Gangways: Emergency end doors only.
Couplers: LU automatic wedgelock.
Control System: Resistor camshaft.
Doors: Sliding.
Maximum Speed: 45 mph.
Train Protection: Tripcock.
Seating Layout: 268 seats, mainly longitudinal but a few 2+2 facing.

DM. Metro-Cammell 1972–74. 40 seats. 27.8 tonnes.
T. Metro-Cammell 1972–74. 36 seats. 18.1 tonnes.
UNDM. Metro-Cammell 1972–74. 40 seats. 26.5 tonnes.
MM. Metro-Cammell 1972–74. 40 seats. 27.8 tonnes.

4-car single-ended units (south end of train)
Cars 4352–4363 are fitted with de-icing equipment.

DM (south)	*T*	*T*	*MM*
3231	4231	4331	3331
3232	4232	4332	3332
3233	4233	4333	3333
3234	4234	4334	3334
3235	4235	4335	3335
3236	4236	4336	3336
3237	4237	4337	3337
3238	4238	4338	3338
3239	4239	4339	3339
3240	4240	4340	3340
3241	4241	4341	3341
3242	4242	4342	3342
3243	4243	4343	3343
3244	4244	4344	3344
3245	4245	4345	3345
3246	4246	4346	3346
3247	4247	4347	3347
3248	4248	4348	3348
3250	4250	4350	3350
3251	4251	4351	3351
3252	4252	4352	3352
3253	4253	4353	3353
3254	4254	4354	3354
3255	4255	4355	3355
3256	4256	4356	3356
3258	4258	4358	3358
3259	4259	4359	3359
3260	4260	4360	3360
3261	4261	4361	3361
3262	4262	4362	3362
3263	4263	4363	3363
3264	4264	4364	3364
3265	4265	4365	3365
3266	4266	4366	3366
3267	4267	4367	3367
3299	4299	4399	3399*

3-car single-ended units (north end of train)

UNDM	*T*	*DM (north)*
3431	4531	3531
3432	4532	3532
3433	4533	3533
3434	4534	3534
3435	4535	3535
3436	4536	3536
3437	4537	3537
3438	4538	3538
3440	4540	3540
3441	4541	3541
3442	4542	3542
3443	4543	3543
3444	4544	3544
3445	4545	3545
3446	4546	3546
3447	4547	3547
3448	4548	3548
3449	4549	3549
3450	4550	3550
3451	4551	3551
3452	4552	3552
3453	4553	3553
3454	4554	3554
3455	4555	3555
3456	4556	3556
3457	4557	3557
3458	4558	3558
3459	4559	3559
3460	4560	3560
3461	4561	3561
3462	4562	3562
3463	4563	3563
3464	4564	3564
3465	4565	3565
3466	4566	3566
3467	4567	3567

▲ 3262/3557 arrives at Warwick Avenue with a service for Elephant & Castle on 10 February 2019.
Robert Pritchard

▼ Interior of Bakerloo Line UNDM 3460, showing the mix of longitudinal and facing seating.
Robert Pritchard

1.2. PICCADILLY LINE

Route: Uxbridge/Heathrow Airport–Cockfosters. **Route length:** 44 miles.
No. of trains: 86 x 6-car (516 cars). **No. of stations:** 53. **First section opened:** 1869.

The Piccadilly Line links Heathrow – Europe's busiest airport, with central London and King's Cross/St Pancras (notably the only main line terminus served by the line), finishing at Cockfosters in north London.

The Piccadilly has its origins in the Great Northern, Piccadilly & Brompton Railway which was formed in 1902 from a merger of the Brompton & Piccadilly Circus Railway and the Great Northern & Strand Railway. Four years later in 1906, the GNP&BR opened its first section of line between Finsbury Park and Hammersmith.

However, this is not the oldest section of the present-day Piccadilly Line: this honour falls to the Hammersmith–Hounslow West section of the present day Heathrow branch, opened by the District Railway in stages between 1869 and 1884. This section was served only by the DR (later to become the District Line) until 1933 when Piccadilly Line trains began to run as far as Northfields on 9 January and Hounslow West on 13 March. District Line trains ceased serving the Hounslow West branch in 1964, since when it has been operated only by the Piccadilly Line.

Also originally operated by the District Railway was the Acton Town–Rayners Lane branch, opened in 1910. In July 1932 the Piccadilly Line took over this route as far as South Harrow, followed in October 1933 by South Harrow–Rayners Lane–Uxbridge (shared with the Metropolitan Line between Rayners Lane and Uxbridge).

The first new-build extension took the Piccadilly Line from Finsbury Park to its northern terminus at Cockfosters, opened in three stages between September 1932 and July 1933. This would be the last extension of the line for over 40 years until the Hounslow West branch was extended to Hatton Cross in July 1975 (requiring the resiting of Hounslow West station), then on to Heathrow Airport in December 1977.

Since then the expansion of Heathrow Airport has seen the addition of Heathrow Terminal 4 station in April 1986 (when Heathrow Central station was renamed Heathrow Terminals 1, 2, 3; now known as Terminals 2 & 3 following the closure of Terminal 1) and Heathrow Terminal 5 in March 2008. Between 1986 and 2008 all Heathrow trains ran in a clockwise direction around the Heathrow Loop, calling at Terminal 4 station first and then Terminals 1, 2, 3. This pattern still applies to trains serving Terminal 4, but Terminal 5 trains serve Terminals 2 & 3 first before continuing to the Terminal 5 terminus.

The only part of the Piccadilly Line to have closed is the Holborn–Aldwych branch, opened in 1907 and closed in 1994, having operated in peak hours only for many years. This branch was considered for extension to Waterloo on a number of occasions but these plans were never realised. Despite its closure, Aldwych station (originally named Strand) became Grade II listed in 2011. The station and branch have long been a popular filming location.

Also closed (although only ever served by the District Railway) was the Hounslow Town spur, a short triangle branch abandoned in 1909 when the present Hounslow East station (named Hounslow Town until 1925) opened.

Apart from Aldwych, the Piccadilly Line has five other closed stations (all located on operational sections of the line):

- Brompton Road, between South Kensington and Knightsbridge (closed 1934);
- Down Street, between Hyde Park Corner and Green Park (closed 1932, but still booked to be used by occasional empty reversing trains);
- Osterley Park & Spring Grove, between Boston Manor and Osterley (closed 1934 and replaced by the present Osterley station);
- Park Royal & Twyford Abbey, between Alperton and Park Royal (closed 1931 and replaced by the present Park Royal station);
- York Road, between King's Cross St Pancras and Caledonian Road (closed 1932).

A noteworthy feature of the Piccadilly Line is "express" running – between Hammersmith and Acton Town, Piccadilly Line trains don't normally serve four stations (Ravenscourt Park, Stamford Brook, Turnham Green and Chiswick Park) that are served by District Line trains (although early morning, late night and Night Tube trains call at Turnham Green).

▲ Piccadilly Line 1973 Stock 134/121 arrives at Arsenal with an Arnos Grove service on 9 February 2019. The station also carries its original name Gillespie Road – it was renamed in 1932 after the nearby football club who played at Highbury Stadium (and from 2006 at the nearby Emirates Stadium). **Robert Pritchard**

▼ 178/111 arrives at Finsbury Park with a Cockfosters service on 31 May 2019. **Robert Pritchard**

The Heathrow branch is almost all in tunnel west of Hounslow West. It is then above ground to Barons Court (as is the entire route from Uxbridge). After Barons Court it is in tunnel to Arnos Grove, and then above ground to Cockfosters apart from a tunnel around a mile long at Southgate.

SERVICES

The basic off-peak Piccadilly Line service pattern consists of:

- Six trains per hour Cockfosters–Heathrow Terminal 5 (via Terminals 2 & 3);
- Six trains per hour Cockfosters–Heathrow Terminal 4 (returning via Terminals 2 & 3);
- Six trains per hour Cockfosters–Rayners Lane, of which three continue to Uxbridge;
- Three trains per hour Arnos Grove–Northfields.

However, it is not quite as simple as that, as there are a number of off-pattern short workings such as Arnos Grove–Heathrow and Ruislip–Cockfosters.

There are a few early morning short workings between Osterley (starting from Northfields depot) and Heathrow, running alternately to Terminal 4 and Terminal 5, with the first train on Mondays–Fridays leaving Osterley at 04.49. In the other direction the first trains leave Terminal 4 at 05.02 and Boston Manor at 05.18, and the first westbound train along the length of the line leaves Cockfosters at 05.09. The last trains covering the core central section of the line reach Rayners Lane at 00.49, Heathrow Terminal 5 at 00.52, Cockfosters at 00.47 and Arnos Grove at 00.41. There are also a few late night short workings, the last of which starts from Acton Town at 00.59 and arrives at Uxbridge at 01.32. On Sunday nights the last trains mostly reach their destination between 00.30 and 01.00.

A Night Tube service operates every 10 minutes on Friday and Saturday nights between Heathrow Terminal 5 and Cockfosters.

ROLLING STOCK AND DEPOTS

All services are formed of 1973 Tube Stock, purpose-built for the Piccadilly Line by Metro-Cammell, Birmingham and refurbished 1993–2001 by Bombardier Prorail at Horbury. These trains were a further development of Victoria Line 1967 Tube Stock and 1972 Tube Stock now used on the Bakerloo Line, but modified with longer car bodies and wider vestibules to provide more luggage space for Heathrow Airport passengers.

There are 86 trains formed of two 3-car half units, not in permanent formations. Some are formed of double-ended DM–T–DM units, but two of these are not normally used together in one train. One double-ended unit is formed of three cars from withdrawn half units. 79 trains are required for the peak service.

The main depots are Northfields and Cockfosters. Trains also stable at Uxbridge, South Harrow, and Arnos Grove.

The existing Piccadilly Line fleet is due to be replaced by 94 new fully-automatic Siemens Inspiro trains by 2026 as part of the TfL's Deep Tube Modernisation Programme. This will see a 60% increase in capacity on the busiest parts of the Piccadilly Line by 2030, with the operation of 27 trains per hour at peak times (from 2026) and then, if an option for seven further train is taken up, to 33 trains per hour (from 2030). There is an option for a further eight trains to enable a 36 trains per hour service to operate. The first of the new trains is due to arrive for testing in summer 2023, entering service between 2024 and 2026.

▲ 143 leads an Uxbridge bound service at Ruislip on 9 September 2017. **Jamie Squibbs**

▼ 867/110 arrives at Covent Garden with a service for Rayners Lane on 22/05/19. 867 is one of the double-ended units, with inter-car fixings on the edges of the cab, noticeable in this photo. Double-ended units rarely operate together. **Robert Pritchard**

1973 TUBE STOCK PICCADILLY LINE

Formation: 6-car; DM–T–UNDM + UNDM–T–DM or DM–T–UNDM + DM–T–DM.
Construction: Aluminium body on steel underframe.
Traction Motors: 4 x DC nose-suspended Brush LT118 of 49 kW (65 hp) per motor car.
Wheel Arrangement: Bo-Bo + 2-2 + Bo-Bo (+) Bo-Bo + 2-2 + Bo-Bo.
Braking: Westcode and rheostatic.
Dimensions: 17.47/17.68 x 2.63 m. **Height**: 2.888 m.
Bogies: Plate frame.
Gangways: Emergency end doors only.
Couplers: LU automatic wedgelock.
Control System: Resistor camshaft.
Doors: Sliding.
Maximum Speed: 45 mph.
Train Protection: Tripcock.
Seating Layout: 228 seats, longitudinal.

DM. Metro-Cammell 1974–77. 38 seats. 29.8 tonnes.
T. Metro-Cammell 1974–77. 38 seats. 20.2 tonnes.
UNDM. Metro-Cammell. 1974–77. 38 seats. 28.5 tonnes.

Cars 606–652 (even numbers only) are fitted with de-icing equipment.

† The T vehicles of sets 868+668+869 and 890+690+891 have been modified to carry rail adhension clearing equipment each autumn. During the autumn they are used as 3-car Sandite trains.

Vehicles 366 and 566 are stored at Northfields Depot.

▲ Interior of Piccadilly Line T car 664. **Alan Yearsley**

3-car single-ended units

DM	*T*	*UNDM*	*DM*	*T*	*UNDM*	*DM*	*T*	*UNDM*
100	500	300	152	552	352	205	605	405
101	501	301	153	553	353	206	606	406
102	502	302	154	554	354	207	607	407
103	503	303	155	555	355	208	608	408
104	504	304	156	556	356	209	609	409
105	505	305	157	557	357	210	610	410
106	506	306	158	558	358	211	611	411
107	507	307	159	559	359	212	612	412
108	508	308	160	560	360	213	613	413
109	509	309	161	561	361	214	614	414
110	510	310	162	562	362	215	615	415
111	511	311	163	563	363	216	616	416
112	512	312	164	564	364	217	617	417
113	513	313	165	565	365	218	618	418
115	515	315	167	567	367	219	619	419
116	516	316	168	568	368	220	620	420
117	517	317	169	569	369	221	621	421
118	518	318	170	570	370	222	622	422
119	519	319	171	571	371	223	623	423
120	520	320	172	572	372	224	624	424
121	521	321	173	573	373	225	625	425
122	522	322	174	574	374	226	626	426
123	523	323	175	575	375	227	627	427
124	524	324	176	576	376	228	628	428
125	525	325	177	577	377	229	629	429
126	526	326	178	578	378	230	630	430
127	527	327	179	579	379	231	631	431
128	528	328	180	580	380	232	632	432
129	529	329	181	581	381	233	633	433
130	530	330	182	582	382	234	634	434
131	531	331	183	583	383	235	635	435
132	532	332	184	584	384	236	636	436
133	533	333	185	585	385	237	637	437
134	534	334	186	586	386	238	638	438
135	535	335	187	587	387	239	639	439
136	536	336	188	588	388	240	640	440
137	537	337	189	589	389	241	641	441
138	538	338	190	590	390	242	642	442
139	539	339	191	591	391	243	643	443
140	540	340	192	592	392	244	644	444
141	541	341	193	593	393	245	645	445
142	542	342	194	594	394	246	646	446
143	543	343	195	595	395	247	647	447
144	544	344	196	596	396	248	648	448
145	545	345	197	597	397	249	649	449
146	546	346	198	598	398	250	650	450
147	547	347	199	599	399	251	651	451
148	548	348	200	600	400	252	652	452
149	549	349	201	601	401	253	653	453
150	550	350	202	602	402	–	566	366 (S)
151	551	351	203	603	403			

3-car double-ended units. * (ex-114 688 889)

DM	*T*	*DM*	*DM*	*T*	*DM*	*DM*	*T*	*DM*
854	654	855	868	668	869 †	882	682	883
856	656	857	870	670	871	884	684	885
858	658	859	872	672	873	886	686	887
860	660	861	874	674	875	890	690	891 †
862	662	863	876	676	877	892	692	893
864	664	865	878	678	879	894	694	895
866	666	867	880	680	881	896	696	897 *

1.3. CENTRAL LINE

Route: West Ruislip/Ealing Broadway–Epping/Hainault. **Route length:** 46 miles.
No. of trains: 85 x 8-car (680 cars). **No. of stations:** 49. **First section opened:** 1900.

The Central Line, as its name suggests, runs across the centre of London, from west to east, serving the shopping districts of Oxford Street, the West End and the City of London. The busiest line on the Underground, it starts from the far suburbs of West Ruislip in the west, finishing at Epping in the east: the 34 mile journey between the two stations is the longest possible on LU, taking between 82 and 87 minutes depending on the time of day.

The Central Line was the third deep level Tube line to open, with the initial section between Queen's Road (now Queensway) and King William Street (now Bank) being opened by the Central London Railway Company in 1900. Until 1907, the CLRC charged a 2d flat fare, hence the "Twopenny Tube" nickname. An extension to Wood Lane, near Shepherd's Bush, opened in 1908 to coincide with the Franco-British Exhibition. This was followed by extensions to Liverpool Street in 1912 and Ealing Broadway in 1920.

As part of the London Passenger Transport Board's 1935–40 New Works Programme the line was extended to Greenford in 1947 and West Ruislip in 1948. This programme also saw the Central Line extended eastwards to Stratford and beyond, taking over the existing Great Eastern suburban lines from Leyton to Epping, Ongar, Hainault and Newbury Park in stages between 1946 and 1949 including a new underground section between Leytonstone and Newbury Park to complete the Hainault Loop. The Epping–Ongar section remained steam-operated until 1957 and closed as part of the LU network in 1994, reopening as a heritage operation under the name of the Epping Ongar Railway in 2004.

The western section from White City to Ealing Broadway and West Ruislip is above ground. From White City the line is in tunnel as far as Stratford, where it runs above ground through Stratford station and then goes back into a tunnel as far as the next station, Leyton, from where the rest of the line is above ground except for the tunnel section of the Hainault Loop between Leytonstone and Newbury Park.

SERVICES

The basic off-peak service pattern sees trains running West Ruislip–Epping and Ealing Broadway–Hainault via Newbury Park (some of which continue to Woodford via the top section of the Hainault Loop). These are supplemented by shorter workings such as White City–Newbury Park and Northolt–Loughton. Peak hours see some non-standard workings such as Ealing Broadway–Epping and West Ruislip–Hainault. A total of 24 trains per hour operate on the core central London section, giving a train every 2–3 minutes. This is increased to 26–30 trains per hour in peak periods.

Operation of 4-car sets on Hainault–Woodford shuttles is planned from the 2020 timetable, releasing one 8-car set for the core routes.

Trains run from around 05.30 to 00.30 on the core section on Mondays–Thursdays, with some short workings between Epping and Loughton until as late as 01.30. The Central Line is one of five Underground lines to operate a Night Tube service on Friday and Saturday nights, with trains running every 20 minutes White City–Hainault via Newbury Park and Ealing Broadway–Loughton, giving a 10-minute service between White City and Leytonstone. On Sunday nights the last trains pass through the central section by around 23.50, reaching the outer ends of the line by around 00.30.

ROLLING STOCK AND DEPOTS

Since its introduction in 1993–95, the 1992 Tube Stock fleet has worked all Central Line services. There are 85 trains formed of four 2-car units in no permanent formation. The 1992 Stock was built by ADtranz (formerly BREL) at Derby. The majority of trains are formed two A+B and two C+B units. A are DM cars with cabs and B and C cabless M cars but with shunting control cabinets. B cars do not have traction pick up shoes. C cars are designated D where fitted with de-icing equipment. All axles on all cars are motored. There are more A than C cars forcing a small number of trains into traffic with boxed-in cabs. All cars received new Siemens bogies from 2011 onwards, replacing the original Kawasaki bogies. 78 trains are required for the peak service.

▲ Central Line 1992 Stock 91319/92168/92156/91177 arrives at Notting Hill Gate with a service for Ealing Broadway on 16 March 2019. **Robert Pritchard**

▼ 91003/92438/92210/91009 arrives at Barkingside with a service for White City on 16 June 2019. **Robert Pritchard**

▲ 91089/92212/92232/91095 arrive at Tottenham Court Road with a service to Loughton on 16 March 2019. **Robert Pritchard**

▼ 91145 arrives at Stratford leading a westbound train on 31 March 2019. **Alan Yearsley**

In 2017 Bombardier won a 7-year contract to supply new AC Mitrac traction motors and traction control equipment for the Central Line fleet, including the provision of ongoing maintenance support. B+D cars 92446+93446 was the first set to be equipped with the new AC motors as a prototype set in 2019. The rest of the fleet is to be fitted with the new traction motors at Acton Works.

The main Central Line depots are at Hainault and Ruislip; trains also stable at White City, Loughton and Woodford.

1992 TUBE STOCK — CENTRAL LINE

Formation: 8-car; DM–M + M–M + M–M + M–DM; some with boxed-in DM formed mid-train.
Construction: Aluminium extrusions.
Traction Motors: 4 x DC fully suspended Brush LT130 of 49 kW (65 hp) per motor car (* new AC Bombardier Mitrac motors, full details awaited).
Wheel Arrangement: Bo-Bo + Bo-Bo (+) Bo-Bo + Bo-Bo (+) Bo-Bo + Bo-Bo (+) Bo-Bo + Bo-Bo.
Braking: Analogue rheostatic and regenerative.
Dimensions: 16.25 x 2.62 m. **Height**: 2.869 m.
Bogies: Kawasaki H frame (original); rebogied Siemens.
Gangways: Emergency end doors only.
Couplers: LU automatic wedgelock.
Control System: GTO chopper.
Doors: Sliding, externally mounted.
Maximum Speed: 53 mph.
Train Protection: Westinghouse coded track circuit ATC.
Seating Layout: 272 seats, longitudinal.

DM. ADtranz 1991–94. 34 seats. 22.5 tonnes.
M(B). ADtranz 1991–94. 34 seats. 20.5 tonnes.
M(C). ADtranz 1991–94. 34 seats. 21.5 tonnes.

A+B 2-car units

DM	*NDM*						
91001	92001	91063	92063	91125	92125	91187	92187
91003	92003	91065	92065	91127	92127	91189	92189
91005	92005	91067	92067	91129	92129	91191	92191
91007	92007	91069	92069	91131	92131	91193	92193
91009	92009	91071	92071	91133	92133	91195	92195
91011	92011	91073	92073	91135	92135	91197	92197
91013	92013	91075	92075	91137	92137	91199	92199
91015	92015	91077	92077	91139	92139	91201	92201
91017	92017	91079	92079	91141	92141	91203	92203
91019	92019	91081	92081	91143	92143	91205	92205
91021	92021	91083	92083	91145	92145	91207	92207
91023	92023	91085	92085	91147	92147	91209	92209
91025	92025	91087	92087	91149	92149	91211	92211
91027	92027	91089	92089	91151	92151	91213	92213
91029	92029	91091	92091	91153	92153	91215	92215
91031	92031	91093	92093	91155	92155	91217	92217
91033	92033	91095	92095	91157	92157	91219	92219
91035	92035	91097	92097	91159	92159	91221	92221
91037	92037	91099	92099	91161	92161	91223	92223
91039	92039	91101	92101	91163	92163	91225	92225
91041	92041	91103	92103	91165	92165	91227	92227
91043	92043	91105	92105	91167	92167	91229	92229
91045	92045	91107	92107	91169	92169	91231	92231
91047	92047	91109	92109	91171	92171	91233	92233
91049	92049	91111	92111	91173	92173	91235	92235
91051	92051	91113	92113	91175	92175	91237	92237
91053	92053	91115	92115	91177	92177	91239	92239
91055	92055	91117	92117	91179	92179	91241	92241
91057	92057	91119	92119	91181	92181	91243	92243
91059	92059	91121	92121	91183	92183	91245	92245
91061	92061	91123	92123	91185	92185	91247	92247

▲ 91203/92252/92034/91281 approach Newbury Park with a service for White City on 16 June 2019. **Robert Pritchard**

▼ Interior of Central Line NDM 92213. **Robert Pritchard**

91249	92249	91275	92275	91301	92301	91327	92327
91251	92251	91277	92277	91303	92303	91329	92329
91253	92253	91279	92279	91305	92305	91331	92331
91255	92255	91281	92281	91307	92307	91333	92333
91257	92257	91283	92283	91309	92309	91335	92335
91259	92259	91285	92285	91311	92311	91337	92337
91261	92261	91287	92287	91313	92313	91339	92339
91263	92263	91289	92289	91315	92315	91341	92341
91265	92265	91291	92291	91317	92317	91343	92343
91267	92267	91293	92293	91319	92319	91345	92345
91269	92269	91295	92295	91321	92321	91347	92347
91271	92271	91297	92297	91323	92323	91349	92349
91273	92273	91299	92299	91325	92325		

B+C 2-car units

NDM	*NDM*						
92002	93002	92070	93070	92136	93136	92202	93202
92004	93004	92072	93072	92138	93138	92204	93204
92006	93006	92074	93074	92140	93140	92206	93206
92008	93008	92076	93076	92142	93142	92208	93208
92010	93010	92078	93078	92144	93144	92210	93210
92012	93012	92080	93080	92146	93146	92212	93212
92014	93014	92082	93082	92148	93148	92214	93214
92016	93016	92084	93084	92150	93150	92216	93216
92018	93018	92086	93086	92152	93152	92218	93218
92020	93020	92088	93088	92154	93154	92220	93220
92022	93022	92090	93090	92156	93156	92222	93222
92024	93024	92092	93092	92158	93158	92224	93224
92026	93026	92094	93094	92160	93160	92226	93226
92028	93028	92096	93096	92162	93162	92228	93228
92030	93030	92098	93098	92164	93164	92230	93230
92032	93032	92100	93100	92166	93166	92232	93232
92034	93034	92102	93102	92168	93168	92234	93234
92036	93036	92104	93104	92170	93170	92236	93236
92038	93038	92106	93106	92172	93172	92238	93238
92040	93040	92108	93108	92174	93174	92240	93240
92042	93042	92110	93110	92176	93176	92242	93242
92044	93044	92112	93112	92178	93178	92244	93244
92046	93046	92114	93114	92180	93180	92246	93246
92048	93048	92116	93116	92182	93182	92248	93248
92050	93050	92118	93118	92184	93184	92250	93250
92052	93052	92120	93120	92186	93186	92252	93252
92054	93054	92122	93122	92188	93188	92254	93254
92056	93056	92124	93124	92190	93190	92256	93256
92058	93058	92126	93126	92192	93192	92258	93258
92060	93060	92128	93128	92194	93194	92260	93260
92062	93062	92130	93130	92196	93196	92262	93262
92064	93064	92132	93132	92198	93198	92264	93264
92066	93066	92134	93134	92200	93200	92266	93266
92068	93068						

B+D 2-car units

NDM	*NDM*						
92402	93402	92418	93418	92434	93434	92450	93450
92404	93404	92420	93420	92436	93436	92452	93452
92406	93406	92422	93422	92438	93438	92454	93454
92408	93408	92424	93424	92440	93440	92456	93456
92410	93410	92426	93426	92442	93442	92458	93458
92412	93412	92428	93428	92444	93444	92460	93460
92414	93414	92430	93430	92446	93446 *	92462	93462
92416	93416	92432	93432	92448	93448	92464	93464

1.4. WATERLOO & CITY LINE

Route: Waterloo–Bank. **Route length:** 1½ miles.
No. of trains: 5 x 4-car (20 cars). **No. of stations:** 2. **First section opened:** 1898.

The Waterloo & City Line connects the main line Waterloo station with the city of London at Bank.

The line was opened in 1898 by the Waterloo & City Railway, a subsidiary of the London & South Western Railway created to cater for main line passengers arriving at Waterloo who wished to reach the City of London, which has been the main purpose of the line since its opening. It is unique in being only a shuttle between two stations, with no intermediate stops, and having no physical connection with either the National Rail network or other Underground lines. The northern terminus of the line was originally referred to as City station by the W&CR and as Bank station by the Central London Railway and the City & South London Railway (the original operators of the Central and Northern Lines respectively), the entire station being named Bank since 1940.

Until April 1994, the W&C was operated by British Railways (and its predecessors before railway nationalisation in 1948). With the privatisation of BR, the line became part of London Underground. Waterloo & City Line DC traction power is still supplied through Network Rail and controlled from Raynes Park control room.

SERVICES

Services operate throughout the day from around 06.15–00.30 on Mondays–Fridays. Off-peak trains operate every 5 minutes, and generally every 3 minutes at peak times. Historically the line has only operated on Saturday mornings, but in recent years this has gradually been extended to an almost all-day Saturday service from around 08.00 to 00.30. There is no Sunday or Bank Holiday service.

ROLLING STOCK AND DEPOTS

Only three types of train have operated on the W&C since its opening. From 1898 until 1940 services were worked by EMUs built by the Jackson & Sharp Company of Wilmington, Delaware in the USA. These were then replaced by units built at Dick Kerr Works, Preston, latterly known as Class 487. Finally, in 1993 these distinctive units were replaced by 4-car 1992 Tube Stock (BR Class 482) of the same basic design as those on the Central Line (BR tagged an order for five 4-car units onto the LU order for Central Line 1992 Stock). The units were originally painted in a version of Network SouthEast blue and white livery.

W&C DM and M cars are equivalent to Central Line A and B cars, but are not equipped for Automatic Train Operation, and are designated E and F. Sets are normally paired as consecutively numbered sets (ie 65501+65502). All five trains are required to be in service at peak times, with three required at off-peak times.

The main depot is at Waterloo (just beyond the end of the platforms) and trains also stable at Bank. The isolation of the line from the rest of the LUL and National Rail networks means that trains have to be lifted off the track by cranes whenever they require maintenance work not able to be carried out on site.

1992 TUBE STOCK WATERLOO & CITY LINE

Formation: 4-car; DM–M + M–DM.
Construction: Aluminium extrusions.
Traction Motors: 4 x DC fully suspended Brush LT130 of 49 kW (65 hp) per motor car.
Wheel Arrangement: Bo-Bo + Bo-Bo (+) Bo-Bo + Bo-Bo.
Braking: Analogue rheostatic and regenerative.
Dimensions: 16.25 x 2.62 m. **Height**: 2.869 m.
Bogies: Kawasaki H frame (original); rebogied Siemens.
Gangways: Emergency end doors only.
Couplers: LU automatic wedgelock.
Control System: GTO chopper.
Doors: Sliding, externally mounted.
Maximum Speed: 53 mph.
Train Protection: Tripcock.
Seating Layout: 136 seats, longitudinal.

DM. ADtranz 1991–94. 34 seats. 22.5 tonnes.
M. ADtranz 1991–94. 34 seats. 21.5 tonnes.

2-car units E (east) + F (middle)

DM	*M*						
65501	67501	65505	67505	65507	67507	65509	67509
65503	67503						

2-car units F (middle) + E (west)

M	*DM*						
67502	65502	67506	65506	67508	65508	67510	65510
67504	65504						

▲ Waterloo & City Line 1992 Stock 65509/510 arrives at Bank with a service from Waterloo on 16 March 2019. **Robert Pritchard**

1.5. NORTHERN LINE

Route: Edgware/High Barnet/Mill Hill East–Morden via Bank/Charing Cross.
Route length: 36 miles.
No. of trains: 106 x 6-car (636 cars). **No. of stations**: 50. **First section opened**: 1890.

The Northern Line runs from three termini in north London – Edgware, High Barnet and Mill Hill East south to Campden Town where it splits into two routes, one serving Bank and London Bridge and the other Charing Cross and Waterloo. The routes rejoin at Kennington and the line continues to Morden – the most southerly station on LU.

The Northern Line has its origins in the City & South London Railway, London's (and the world's) first deep level tube railway opened between Stockwell and King William Street in 1890. By the turn of the 20th century the C&SL had been extended northwards to Moorgate and southwards to Clapham Common. This was followed by extensions to Angel in 1901 and Euston in 1907.

Also in 1907 the Charing Cross, Euston & Hampstead Railway, having been bought by American financier Charles Yerkes in 1900, opened its own line from Charing Cross to Golders Green with a branch from Camden Town to Archway. In 1913 the Yerkes Group purchased the C&SLR, and further extensions were built between the 1920s and early 1940s. By 1924 the Golders Green branch had reached its present terminus at Edgware, and in 1926 the original CCE&H line was extended southwards to Morden. Finally, the Archway branch was extended to East Finchley in 1939 and High Barnet in 1940, with a short branch being built to Mill Hill East in 1941 (this having been planned to be extended to Edgware).

The name "Northern Line" was first adopted in 1937, ironically after the never realised "Northern Heights" plan to extend the line beyond Edgware to Brockley Hill, Elstree and Bushey Heath. This scheme was included in the 1935–40 New Works Programme, but was abandoned after World War II when the area north of Edgware became part of the Metropolitan Green Belt.

With the formation of the London Passenger Transport Board in 1933, the Great Northern & City Railway between Moorgate and Finsbury Park was made part of what would become the Northern Line. As part of the Northern Heights scheme this line was to have been extended beyond Finsbury Park over the now closed Alexandra Palace branch and via Mill Hill East to Edgware, but these plans never came to fruition, and the Northern City Line remained a self-contained operation until its transfer to British Rail in 1975.

Thus, apart from the disposal of the Northern City Line, the route of the Northern Line has remained unchanged since World War II. Today the Northern Line is the second busiest line on the Underground. A new branch from Kennington to Battersea is planned to be completed by 2020, the first extension for almost 80 years.

Around 70% of the Northern Line is in tunnel. The northern branches of Edgware (to just beyond Golders Green) and High Barnet (to just beyond East Finchley) plus the Mill Hill East branch are above ground, with the rest, apart from the approach to Morden station (and the link line to Morden depot) being in tunnel. The 17-mile tunnel from East Finchley to Morden via Bank is the longest on LU, while 36 of the 50 stations on the Northern Line are underground (the most on any line). It is also worth noting that the line includes both the deepest (Hampstead station with a 55.2 m lift shaft) and the highest point on the LU network above ground level (Dollis Brook Viaduct: 18 m (60 ft)).

SERVICES

For many years all three northern termini (Edgware Road, High Barnet and Mill Hill East) were served by trains from Morden running both via Charing Cross and via Bank. In recent years the service pattern has been simplified so almost all trains via Charing Cross start and finish at Kennington. Only a small number of trains in peak hours and in the early morning and late evening run through to and from Morden via Charing Cross.

The basic off-peak service consists of ten trains per hour on each of the main flows: Edgware–Kennington via Charing Cross, Edgware–Morden via Bank, High Barnet–Kennington via Charing Cross, and High Barnet–Morden via Bank. Trains that terminate at Kennington use the Kennington loop to turn and then return north. This timetable means that from both Edgware and High Barnet trains run alternately to Kennington via Charing Cross and to Morden via Bank, and the entire line has a 20 tph service except for the Mill Hill East branch which operates as a 4 tph shuttle to and from Finchley Central. Mill Hill East, the least used station on the Northern Line, does have a few through trains to Morden and Kennington in the peaks.

▲ Northern Line 1995 Stock 51583/582 arrives at West Finchley with a High Barnet–Morden service on 24 June 2019. **David Palmer**

▼ Interior of Northern Line DM 51551 showing the tip-up seats nearest the camera. **Alan Yearsley**

Services start at around 05.30 on Mondays–Fridays and finish by 00.30 on Monday–Thursday nights and midnight on Sunday nights. A Night Tube service operates on Friday and Saturday nights, with trains about every 8 minutes between Morden and Camden Town via Charing Cross alternately serving Edgware and High Barnet. Only the Mill Hill East branch and the route via Bank are not served by the Night Tube.

From September 2021 services are due to start operating on the Kennington–Battersea extension.

ROLLING STOCK AND DEPOTS

All services are formed of 1995 Tube Stock trains built by GEC Alsthom, Birmingham which first entered service in June 1998 and have worked the Northern Line since the late 1990s, the last Mark 1 1972 Stock trains having been withdrawn in 1999 and the last 1959 Stock in 2000. There are 106 trains (the largest number on any LU line), formed of two 3-car half units in semi-permanent formations. Despite its designation, 1995 Stock was developed after 1996 Stock, with later IGBT traction power electronics replacing the earlier GTO system and other technical differences. Unlike all other LU trains, 1995 Stock is owned by HSBC/RBS (103 trains) or Alstom (three trains) and leased back to LU. 96 trains are required for traffic at peak times, and 80 at off-peak times and at weekends.

The main depots are at Morden and Golders Green. Morden supplies 38 trains into traffic on weekday mornings – the most from any London Underground depot. Trains also stable at Edgware, High Barnet and Highgate.

1995 TUBE STOCK NORTHERN LINE

Formation: 6-car; DM–T–UNDM + UNDM–T–DM.
Construction: Aluminium extrusions.
Traction Motors: 4 x AC fully suspended GEC G355AZ of 85 kW (114 hp) per motor car.
Wheel Arrangement: Bo-Bo + 2-2 + Bo-Bo (+) Bo-Bo + 2-2 + Bo-Bo.
Braking: Analogue rheostatic and regenerative.
Dimensions: 17.77 x 2.63 m. **Height:** 2.875 m.
Bogies: Flexible box frame.
Gangways: Emergency end doors only.
Couplers: LU automatic wedgelock.
Control System: IGBT.
Doors: Sliding, externally mounted.
Maximum Speed: 45 mph.
Train Protection: Thales Seltrac TBTC.
Seating Layout: 200 seats, longitudinal + 48 tip-ups.

DM. GEC Alsthom 1996–2000. 32 seats + 8 tip-up. 29.4 tonnes.
T. GEC Alsthom 1996–2000. 34 seats + 8 tip-up. 21.5 tonnes.
UNDM. GEC Alsthom 1996–2000. 34 seats + 8 tip-up. 27.9 tonnes.

Cars 52701–726 are fitted with de-icing equipment.

3-car single-ended units

DM	*T*	*UNDM*
51501	52501	53501
51503	52503	53503
51505	52505	53505
51507	52507	53507
51509	52509	53509
51511	52511	53511
51513	52513	53513
51515	52515	53515
51517	52517	53517
51519	52519	53519
51521	52521	53521
51523	52523	53523
51525	52525	53525
51527	52527	53527
51529	52529	53529
51531	52531	53531
51533	52533	53533
51535	52535	53535
51537	52537	53537
51539	52539	53539
51541	52541	53541
51543	52543	53543
51545	52545	53545
51547	52547	53547
51549	52549	53549
51551	52551	53551
51553	52553	53553
51555	52555	53555
51557	52557	53557
51559	52559	53559
51561	52561	53561
51563	52563	53563
51565	52565	53565
51567	52567	53567
51569	52569	53569
51571	52571	53571
51573	52573	53573
51575	52575	53575
51577	52577	53577
51579	52579	53579
51581	52581	53581
51583	52583	53583
51585	52585	53585
51587	52587	53587
51589	52589	53589
51591	52591	53591
51593	52593	53593
51595	52595	53595
51597	52597	53597
51599	52599	53599
51601	52601	53601

51603	52603	53603
51605	52605	53605
51607	52607	53607
51609	52609	53609
51611	52611	53611
51613	52613	53613
51615	52615	53615
51617	52617	53617
51619	52619	53619
51621	52621	53621
51623	52623	53623
51625	52625	53625
51627	52627	53627
51629	52629	53629
51631	52631	53631
51633	52633	53633
51635	52635	53635
51637	52637	53637
51639	52639	53639
51641	52641	53641
51643	52643	53643
51645	52645	53645
51647	52647	53647
51649	52649	53649
51651	52651	53651
51653	52653	53653
51655	52655	53655
51657	52657	53657
51659	52659	53659
51661	52661	53661
51663	52663	53663
51665	52665	53665
51667	52667	53667
51669	52669	53669
51671	52671	53671
51673	52673	53673
51675	52675	53675
51677	52677	53677
51679	52679	53679
51681	52681	53681
51683	52683	53683
51685	52685	53685

3-car single-ended units

UNDM	*T*	*DM*
53502	52502	51502
53504	52504	51504
53506	52506	51506
53508	52508	51508
53510	52510	51510
53512	52512	51512
53514	52514	51514
53516	52516	51516
53518	52518	51518
53520	52520	51520
53522	52522	51522
53524	52524	51524
53526	52526	51526
53528	52528	51528
53530	52530	51530
53532	52532	51532
53534	52534	51534
53536	52536	51536
53538	52538	51538
53540	52540	51540
53542	52542	51542
53544	52544	51544
53546	52546	51546
53548	52548	51548
53550	52550	51550
53552	52552	51552
53554	52554	51554
53556	52556	51556
53558	52558	51558
53560	52560	51560
53562	52562	51562
53564	52564	51564
53566	52566	51566
53568	52568	51568
53570	52570	51570
53572	52572	51572
53574	52574	51574
53576	52576	51576
53578	52578	51578
53580	52580	51580
53582	52582	51582
53584	52584	51584
53586	52586	51586
53588	52588	51588
53590	52590	51590
53592	52592	51592
53594	52594	51594
53596	52596	51596
53598	52598	51598
53600	52600	51600
53602	52602	51602
53604	52604	51604
53606	52606	51606
53608	52608	51608
53610	52610	51610
53612	52612	51612
53614	52614	51614
53616	52616	51616
53618	52618	51618
53620	52620	51620
53622	52622	51622
53624	52624	51624
53626	52626	51626
53628	52628	51628
53630	52630	51630
53632	52632	51632
53634	52634	51634
53636	52636	51636
53638	52638	51638
53640	52640	51640
53642	52642	51642
53644	52644	51644
53646	52646	51646
53648	52648	51648
53650	52650	51650
53652	52652	51652
53654	52654	51654
53656	52656	51656
53658	52658	51658
53660	52660	51660
53662	52662	51662
53664	52664	51664
53666	52666	51666
53668	52668	51668
53670	52670	51670
53672	52672	51672
53674	52674	51674
53676	52676	51676
53678	52678	51678
53680	52680	51680
53682	52682	51682
53684	52684	51684
53686	52686	51686

3-car single-ended units with de-icing equipment

DM	*T*	*UNDM*
51701	52701	53701
51703	52703	53703
51705	52705	53705
51707	52707	53707
51709	52709	53709
51711	52711	53711
51713	52713	53713
51715	52715	53715
51717	52717	53717
51719	52719	53719
51721	52721	53721
51723	52723	53723
51725	52725	53725

3-car single-ended units with de-icing equipment

UNDM	*T*	*DM*
53702	52702	51702
53704	52704	51704
53706	52706	51706
53708	52708	51708
53710	52710	51710
53712	52712	51712
53714	52714	51714
53716	52716	51716
53718	52718	51718
53720	52720	51720
53722	52722	51722
53724	52724	51724
53726	52726	51726

NORTHERN LINE FORMATIONS

Sets of 1995 Stock operate in semi-permanently-coupled formations, although there have been a few changes since the publication of the first edition of this book. At the time of writing formations were as follows:

51501/701
51502/503
51504/505
51506/507
51508/702
51509/510
51511/512
51513/514
51515/703
51516/517
51518/519
51520/704
51521/522
52523/524
51525/717
51526/527
51528/529
51530/557
51531/556
51532/706
51533/534
51535/536
51537/538
51539/707
51540/541
51542/543
51544/545
51546/708
51547/548
51549/550
51551/711
51552/567
51553/709
51554/555
51558/559
51560/710
51561/562
51563/564
51565/566
51568/569
51570/571
51572/573
51574/712
51575/576
51577/578
51579/580
51581/713
51582/583
51584/585
51586/587
51588/714
51589/590
51591/592
51593/594
51595/715
51596/597
51598/599
51600/601
51602/716
51603/604
51605/606
51607/608
51609/705
51610/611
51612/613
51614/615
51616/718
51617/618
51619/620
51621/622
51623/719
51624/625
51626/627
51628/629
51630/720
51631/632
51633/634
51635/636
51637/721
51638/639
51640/641
51642/643
51644/722
51645/646
51647/648
51649/650
51651/723
51652/653
51654/655
51656/657
51658/724
51659/660
51661/662
51663/664
51665/725
51666/667
51668/669
51670/671
51672/726
51673/674
51675/676
51677/678
51679/680
51681/682
51683/684
51685/686

▲ The majority of Northern Line trains run in consecutively numbered sets, but there are exceptions, particularly to include the cars with de-icing equipment (in the 517xx series). 51513/514 pauses at Highgate, which has long platforms as it was originally built to take 9-car trains, with a service to Morden via Bank on 6 April 2019. **Robert Pritchard**

1.6. JUBILEE LINE

Route: Stanmore–Stratford. **Route length:** 22½ miles.
No. of trains: 63 x 7-car (441 cars). **No. of stations:** 27. **First section opened:** 1932.

The Jubilee Line, the newest of the deep Tube lines, runs from Stanmore in north London through to Waterloo before heading east to Canary Wharf and Stratford.

The northern section of the Jubilee Line, between Baker Street and Stanmore, is the oldest stretch of the line and has its origins in the Metropolitan Railway's branch from Wembley Park to Stanmore, opened in 1932. For the first seven years of its life, this line simply operated as a branch of the Metropolitan. From 1939 Baker Street–Stanmore became a branch of the Bakerloo Line from Elephant & Castle, with this arrangement continuing until the opening of the new section between Baker Street and Charing Cross via Bond Street in 1979 when the "Jubilee Line" name came into being and took over the Stanmore branch.

When the extension south from Baker Street was being planned in the 1970s, it was envisaged that the line would later be extended beyond Charing Cross into South-East London and the Docklands, following a route beneath Fleet Street. Because of this, it was to be named the Fleet Line. In the event, this scheme as it was then proposed never came to fruition, and in 1977 the name Jubilee Line was chosen instead to mark Queen Elizabeth II's Silver Jubilee that year.

In the 1980s and 1990s the regeneration of London's Docklands led to a revival of plans to extend the line eastwards. These eventually resulted in the Jubilee Line Extension scheme which took the line to a new terminus at Stratford. Opening in 1999, the new section left the existing line between Green Park and its Charing Cross terminus, serving Westminster, Waterloo, London Bridge and Canary Wharf. An interchange with the East London Railway was built at Canada Water, and a new station and bus interchange was also provided at North Greenwich to serve the Millennium Dome (now the O2 Arena). The eastern extension opened in three stages starting with Stratford–North Greenwich in May 1999. Completion of this extension in November 1999 led to the closure of the line's previous terminus at Charing Cross to regular services, but it is still occasionally used for diversions during disruption and for filming.

Platform edge doors, which line up with the train doors, are fitted at all tunnel stations on the eastern extension from Westminster to North Greenwich. These are a novel feature on the London Underground but have been used on the metros in Singapore and Lille, France since the 1980s. They are also being used on the new "Elizabeth Line" (Crossrail) route through London.

The Stanmore–Finchley Road and Canning Town–Stratford sections are above ground, and Finchley Road–Canning Town is in tunnel.

SERVICES

The basic off-peak service consists of 23–24 trains per hour on the core section between Willesden Green and North Greenwich, giving a 2–3 minute service frequency. Of these, between ten and 16 trains cover the entire line between Stanmore and Stratford. The rest are short workings covering the centre section of the Line, such as Willesden Green–Stratford, Wembley Park–North Greenwich or Stanmore–West Ham. The peak service sees 30 trains per hour in both directions on the middle section of the Line.

Trains start running at around 05.30 on Mondays–Fridays, finishing between 00.45 and 01.00 on Monday–Thursday nights and around midnight on Sunday nights. After the last train covering the entire line leaves Stanmore at 23.26 on Sundays, there are then two short workings to Neasden. A Night Tube service operates every 10 minutes along the entire line on Friday and Saturday nights.

ROLLING STOCK AND DEPOTS

Since 1998 the Jubilee Line has been operated by trains of 1996 Stock, built by GEC Alsthom-Metro-Cammell (now Alstom), Birmingham to replace the short-lived 1983 Stock. There are 63 7-car trains formed of one north end 3-car plus one south end 4-car unit. The 1996 Stock is of a similar design to the 1995 Stock used on the Northern Line. Despite its designation the 1996 Stock was actually developed before 1995 Stock, 1996 Stock having older GTO traction systems and other detail differences.

▲ Jubilee Line 1996 Stock 96013 leads a southbound train for Stratford past Neasden depot on 8 November 2018. **Jamie Squibbs**

▼ 96051/100 arrives at Swiss Cottage with a service for Stratford on 16 March 2019. **Robert Pritchard**

Originally 59 6-car trains were built, all augmented to 7-cars in 2005. At the same time four completely new 7-car trains were built. The additional units and seventh cars were built by CAF, Zaragoza, Spain for Alstom, retaining the original design traction packs. 58 trains are required for the peak service.

The main depot is at Stratford Market, between Stratford and West Ham stations. This was constructed as part of the Stratford extension – before that Jubilee Line trains had been based at Neasden. Today trains also still stable at Neasden depot (which is shared with the Metropolitan Line) and at Stanmore.

1996 TUBE STOCK — JUBILEE LINE

Formation: 7-car; DM–T–T–UNDM + UNDM–T–DM.
Construction: Aluminium extrusions.
Traction Motors: 4 x AC fully suspended GEC LT200 of 90 kW (120 hp) per motor car.
Wheel Arrangement: Bo-Bo + 2-2 + Bo-Bo (+) Bo-Bo + 2-2 + 2-2 + Bo-Bo.
Braking: Analogue rheostatic and regenerative.
Dimensions: 17.77 x 2.63 m. **Height:** 2.875 m.
Bogies: Flexible box frame.
Gangways: Emergency end doors only.
Couplers: LU automatic wedgelock.
Control System: GTO.
Doors: Sliding, externally mounted.
Maximum Speed: 62 mph.
Train Protection: Thales Seltrac TBTC.
Seating Layout: 238 seats, longitudinal.

DM. GEC Alsthom 1996–98 or *CAF Zaragoza, Spain 2005. 32 seats. 30.0 tonnes.
T. GEC Alsthom 1996–98 or *CAF Zaragoza, Spain 2005. 34 seats. 20.9 tonnes.
UNDM. GEC Alsthom 1996–98 or *CAF Zaragoza, Spain 2005. 34 seats. 27.1 tonnes.

4-car single-ended units

DM (east)	*T*	*T**	*UNDM*
96001	96201	96601*	96401
96003	96203	96603*	96403
96005	96205	96605*	96405
96007	96207	96607*	96407
96009	96209	96609*	96409
96011	96211	96611*	96411
96013	96213	96613*	96413
96015	96215	96615*	96415
96017	96217	96617*	96417
96019	96219	96619*	96419
96021	96221	96621*	96421
96023	96223	96623*	96423
96025	96225	96625*	96425
96027	96227	96627*	96427
96029	96229	96629*	96429
96031	96231	96631*	96431
96033	96233	96633*	96433
96035	96235	96635*	96435
96037	96237	96637*	96437
96039	96239	96639*	96439
96041	96241	96641*	96441
96043	96243	96643*	96443
96045	96245	96645*	96445
96047	96247	96647*	96447
96049	96249	96649*	96449
96051	96251	96651*	96451
96053	96253	96653*	96453
96055	96255	96655*	96455
96057	96257	96657*	96457
96059	96259	96659*	96459
96061	96261	96661*	96461
96063	96263	96663*	96463
96065	96265	96665*	96465
96067	96267	96667*	96467
96069	96269	96669*	96469
96071	96271	96671*	96471
96073	96273	96673*	96473
96075	96275	96675*	96475
96077	96277	96677*	96477
96079	96279	96679*	96479
96081	96281	96681*	96481
96083	96283	96683*	96483
96085	96285	96685*	96485
96087	96287	96687*	96487
96089	96289	96689*	96489
96091	96291	96691*	96491
96093	96293	96693*	96493
96095	96295	96695*	96495
96097	96297	96697*	96497
96099	96299	96699*	96499
96101	96301	96701*	96501
96103	96303	96703*	96503
96105	96305	96705*	96505
96107	96307	96707*	96507
96109	96309	96709*	96509
96111	96311	96711*	96511
96113	96313	96713*	96513
96115	96315	96715*	96515
96117	96317	96717*	96517
96119*	96319*	96719*	96519*
96121*	96321*	96721*	96521*
96123*	96323*	96723*	96523*
96125*	96325*	96725*	96525*

3-car single-ended units

Cars 96880–918 are fitted with de-icing equipment.

UNDM	*T*	*DM (west)*
96402	96202	96002
96404	96204	96004
96406	96206	96006
96408	96208	96008
96410	96210	96010
96412	96212	96012
96414	96214	96014
96416	96216	96016
96418	96218	96018
96420	96220	96020
96422	96222	96022
96424	96224	96024
96426	96226	96026
96428	96228	96028
96430	96230	96030
96432	96232	96032
96434	96234	96034
96436	96236	96036
96438	96238	96038
96440	96240	96040
96442	96242	96042
96444	96244	96044
96446	96246	96046
96448	96248	96048
96450	96250	96050
96452	96252	96052
96454	96254	96054
96456	96256	96056
96458	96258	96058
96460	96260	96060
96462	96262	96062
96464	96264	96064
96466	96266	96066
96468	96268	96068
96470	96270	96070
96472	96272	96072
96474	96274	96074
96476	96276	96076
96478	96278	96078
96480	96880	96080
96482	96882	96082
96484	96884	96084
96486	96886	96086
96488	96888	96088
96490	96890	96090
96492	96892	96092
96494	96894	96094
96496	96896	96096
96498	96898	96098
96500	96900	96100
96502	96902	96102
96504	96904	96104
96506	96906	96106
96508	96908	96108
96510	96910	96110
96512	96912	96112
96514	96914	96114
96516	96916	96116
96518	96918	96118
96520*	96320*	96120*
96522*	96322*	96122*
96524*	96324*	96124*
96526*	96326*	96126*

▲ Interior of Jubilee Line trailer 96230. **Alan Yearsley**

1.7. VICTORIA LINE

Route: Walthamstow Central–Brixton. **Route Length:** 13½ miles.
No. of trains: 47 x 8-car (376 cars). **No. of stations:** 16. **First section opened:** 1968.

The Victoria Line, running from Walthamstow Central in north-west London to Brixton in south London, is unique in at least three ways: it is the only line to have been built completely from scratch since World War II, the only line to have used Automatic Train Operation since its opening, and the only line (apart from the Waterloo & City) to run entirely in tunnels with no above-ground sections used by passenger-carrying trains. It was originally conceived as an express Tube line from Victoria to Finsbury Park in the late 1930s which would then have taken over a number of LNER lines from Liverpool Street.

After World War II this scheme evolved into a Victoria–Walthamstow line which would eventually materialise with the opening of the Walthamstow Central–Highbury & Islington section in September 1968, continuing to Warren Street in December of that year and finally to Victoria in March 1969.

In the immediate post-war period London County Council's County of London Plan had envisaged a southern extension of the Victoria Line from Victoria to Brixton, Streatham, Norbury and East Croydon. In the event only the Victoria–Brixton section saw the light of day, opening in July 1971 with an intermediate station at Vauxhall. A second station at Pimlico followed in September 1972, the only station on the line not to have interchange with another Underground line or with National Rail or London Overground services.

▲ Victoria Line 2009 Stock 11024/023 arrives at Finsbury Park with a service for Walthamstow Central on 9 February 2019. **Robert Pritchard**

SERVICES

Most trains cover the entire line, but there are a few short workings that start or terminate at Seven Sisters particularly in the early morning and late evening when trains start and finish their day's work at Seven Sisters depot. There is also one early morning train that starts at Victoria, but otherwise all trains start or terminate at Brixton. The basic off-peak service pattern sees a train every 2–3 minutes. This is increased to less than every 2 minutes during peak hours, amounting to 33 trains per hour and giving the most intensive train service on any railway in the UK. The Victoria Line is the most intensively used line in terms of the average number of journeys per mile.

The Victoria Line is one of five Underground lines to operate a Night Tube service on Friday and Saturday nights, with trains running every 10 minutes on the entire line. Otherwise, trains run from around 05.30–00.10 Mondays–Thursdays, finishing at around midnight on Sunday nights.

ROLLING STOCK AND DEPOTS

All services are formed of 2009 Tube Stock, which entered service between 2009 and 2011 replacing the 1967 Stock which had been used since the line opened. The 2009 Stock trains are 3 m longer than their predecessors, making them the second longest trains operated by London Underground (at 133.28 m over couplers) after the S8 Stock (133.47 m). There are 47 8-car trains built by Bombardier, Derby which operate in fixed formations. Early pre-production cars forming complete test trains were built for development only; they were never delivered to LU and are believed to have since been broken up. 41 trains are required for the peak service.

The main depot is at Northumberland Park, which is reached via a branch diverging from the Victoria line to the north of Seven Sisters station. The depot is the only part of the line that is above ground, and is located adjacent to Northumberland Park station on the London Liverpool Street–Cambridge main line. Trains also stable at Walthamstow Central, Victoria and Brixton.

2009 TUBE STOCK — VICTORIA LINE

Formation: 8-car; DM–T–NDM–UNDM + UNDM–NDM–T–DM.
Construction: Bolted and welded aluminium extrusions.
Traction Motors: 4 x AC fully suspended Bombardier Mitrac of 75 kW (100 hp) per motor car.
Wheel Arrangement: Bo-Bo + 2-2 + Bo-Bo + Bo-Bo (+) Bo-Bo + Bo-Bo + 2-2 + Bo-Bo.
Braking: Analogue rheostatic and regenerative.
Dimensions: 16.60/16.35 x 2.616 m. **Height:** 2.883 m.
Bogies: Flexible frame (Bombardier).
Gangways: Emergency end doors only.
Couplers: LU automatic wedgelock with pneumatic connections.
Control System: IGBT.
Doors: Sliding, externally mounted.
Maximum Speed: 50 mph.
Train Protection: Westinghouse DTGR.
Seating Layout: 252 seats, longitudinal + 36 tip-ups.

DM. Bombardier 2009–11. 32 seats + 4 tip-ups. 27.1 tonnes.
T. Bombardier 2009–11. 32 seats + 4 tip-ups. 21.6 tonnes.
NDM. Bombardier 2009–11. 32 seats + 4 tip-ups. 23.8 tonnes.
UNDM. Bombardier 2009–11. 30 + 6 tip-ups. 25.8 tonnes.

8-car units

DM (south)	*T*	*NDM*	*UNDM*	*UNDM*	*NDM*	*T*	*DM (north)*
11001	12001	13001	14001	14002	13002	12002	11002
11003	12003	13003	14003	14004	13004	12004	11004
11005	12005	13005	14005	14006	13006	12006	11006
11007	12007	13007	14007	14008	13008	12008	11008
11009	12009	13009	14009	14010	13010	12010	11010
11011	12011	13011	14011	14012	13012	12012	11012
11013	12013	13013	14013	14014	13014	12014	11014
11015	12015	13015	14015	14016	13016	12016	11016
11017	12017	13017	14017	14018	13018	12018	11018
11019	12019	13019	14019	14020	13020	12020	11020

▲ 11032/031 arrives at Seven Sisters with a service for Walthamstow Central on 9 February 2019. **Robert Pritchard**

▼ Interior of Victoria Line DM 11032. **Robert Pritchard**

11021	12021	13021	14021	14022	13022	12022	11022
11023	12023	13023	14023	14024	13024	12024	11024
11025	12025	13025	14025	14026	13026	12026	11026
11027	12027	13027	14027	14028	13028	12028	11028
11029	12029	13029	14029	14030	13030	12030	11030
11031	12031	13031	14031	14032	13032	12032	11032
11033	12033	13033	14033	14034	13034	12034	11034
11035	12035	13035	14035	14036	13036	12036	11036
11037	12037	13037	14037	14038	13038	12038	11038
11039	12039	13039	14039	14040	13040	12040	11040
11041	12041	13041	14041	14042	13042	12042	11042
11043	12043	13043	14043	14044	13044	12044	11044
11045	12045	13045	14045	14046	13046	12046	11046
11047	12047	13047	14047	14048	13048	12048	11048
11049	12049	13049	14049	14050	13050	12050	11050
11051	12051	13051	14051	14052	13052	12052	11052
11053	12053	13053	14053	14054	13054	12054	11054
11055	12055	13055	14055	14056	13056	12056	11056
11057	12057	13057	14057	14058	13058	12058	11058
11059	12059	13059	14059	14060	13060	12060	11060
11061	12061	13061	14061	14062	13062	12062	11062
11063	12063	13063	14063	14064	13064	12064	11064
11065	12065	13065	14065	14066	13066	12066	11066
11067	12067	13067	14067	14068	13068	12068	11068
11069	12069	13069	14069	14070	13070	12070	11070
11071	12071	13071	14071	14072	13072	12072	11072
11073	12073	13073	14073	14074	13074	12074	11074
11075	12075	13075	14075	14076	13076	12076	11076
11077	12077	13077	14077	14078	13078	12078	11078
11079	12079	13079	14079	14080	13080	12080	11080
11081	12081	13081	14081	14082	13082	12082	11082
11083	12083	13083	14083	14084	13084	12084	11084
11085	12085	13085	14085	14086	13086	12086	11086
11087	12087	13087	14087	14088	13088	12088	11088
11089	12089	13089	14089	14090	13090	12090	11090
11091	12091	13091	14091	14092	13092	12092	11092
11093	12093	13093	14093	14094	13094	12094	11094

1.8. SUB-SURFACE LINES

The four Sub-Surface Lines together make up almost 40% of the London Underground network. 192 new S Stock trains built between 2008 and 2015 by Bombardier, Derby revolutionised the Sub-Surface Lines, replacing all of the ageing A60, C and D78 Stock trains and creating a uniform fleet across all SSLs for the first time in the history of the London Underground (the only difference being that the S7 Stock is formed of 7-car trains with all longitudinal seating while the S8 Stock is made up of 8-cars with some facing bays of seats). This order for 1404 vehicles is the largest ever single rolling stock order placed for the same design of UK rolling stock. They are the first London Underground trains to be fitted with air-conditioning.

1.8.1. CIRCLE LINE

Route: Hammersmith/Edgware Road–Liverpool Street–Victoria–Edgware Road (circle).
Route Length: 17 miles.
No. of trains: 133 x 7-car (931 cars)*. **No. of stations:** 36. **First section opened:** 1863.
* Fleet shared with the District and Hammersmith & City Lines.

The Circle Line no longer operates as a true "circle", but more a "teacup" arrangement. As described below, trains run from Hammersmith to Edgware Road, then in a clockwise circle around the centre of London before reaching Edgware Road again: they then return to Hammersmith in an anti-clockwise direction.

The line has its origins in the Metropolitan Railway's original Underground line between Paddington and Farringdon, opened in 1863 (see Metropolitan and Hammersmith & City lines). In the same year a parliamentary select committee report recommended an "inner circle" connecting the existing MR line with London's main line rail termini at Blackfriars, Cannon Street, Charing Cross and Victoria. 1864 saw the formation of the Metropolitan District Railway to open a line from South Kensington to Tower Hill (see District Line).

After being built in stages, enough of the line had been completed to allow the introduction of an Inner Circle service between Moorgate and Mansion House via High Street Kensington on 3 July 1871. 11 years later, in 1882, the Tower Hill–Aldgate section opened, followed by Mansion House–Tower Hill in 1884 when a circular service around the entire circle started, initially steam-hauled until electrification was completed in 1905.

The Circle Line appeared as a unique line on the LU maps from 1949, coloured yellow. The line is unique in that its entire route (apart from the very short sections between Aldgate and Tower Hill and the curve between High Street Kensington and Gloucester Road) is shared with other lines: the Hammersmith & City Line between Hammersmith and Edgware Road and on the top part of the circle as far as Liverpool Street (plus the Metropolitan Line between Baker Street and Aldgate), and the District Line between Gloucester Road and Tower Hill. There are no stations served only by the Circle Line.

Whereas the Hammersmith branch is almost all above ground (see Hammersmith & City Line) the entire route of the original Circle Line is in tunnel apart from a few short gaps in the tunnels including those at Notting Hill Gate, Paddington (District/Circle Line platforms), Edgware Road, Farringdon and Barbican stations.

SERVICES

From 1884 until December 2009, the Circle Line operated as a continuous circle without a fixed terminus. This was then changed so that Circle Line trains start from Hammersmith, following the route of the Hammersmith & City Line, then joining the existing Circle Line at Edgware Road where they terminate after doing a complete circuit. They then return round the circle in an anticlockwise direction before terminating back at Hammersmith. This pattern of operation was introduced to make it easier to recover the service more quickly in case of disruption on any part of the route, but means that passengers making journeys such as Gloucester Road–Baker Street or Notting Hill Gate–Euston Square now always have to change at Edgware Road. Service frequencies between Hammersmith and Edgware Road were also almost doubled as a result of this change. 33 trains are required for the peak Circle and Hammersmith & City Line services.

The basic off-peak Circle Line service pattern sees six trains per hour in each direction, approximately every 10 minutes, but these are complemented by District, Metropolitan and Hammersmith & City Line trains along almost the entire route. Services run from around 05.30–00.30 Mondays–Saturdays and 07.00–midnight on Sundays.

Some early morning trains start from Barking and run as far as Tower Hill following the route of the District Line, then continue to Edgware Road as Circle Line trains. In the other direction there are two late night trains (one on Sundays) that start from Aldgate, run a full clockwise circuit of the Circle Line to Liverpool Street and then continue to Barking.

1.8.2. DISTRICT LINE

Route: Ealing Broadway/Richmond/Wimbledon/Kensington Olympia–Edgware Road/Upminster.
Route Length: 40 miles.
No. of trains: 133 x 7-car (931 cars)*. **No. of stations:** 60. **First section opened:** 1868.
* Fleet shared with the Circle and Hammersmith & City Lines.

The District Line is a complex operation, running from termini in west and south-west London (Ealing Broadway, Richmond and Wimbledon) to Edgware Road or across the south of London to Barking or Upminster, the easternmost station on LU. The District Line has 60 stations, the highest of any LU line, although more than half of them are shared with other lines.

Only a year after the formation of the Metropolitan Railway, 1864 saw the establishment of the Metropolitan District Railway (soon shortened to the District Railway to avoid confusion) to which the present-day District Line can trace its origins. The first section of what is now the District Line opened between Paddington, Gloucester Road and Westminster in 1868 (although High Street Kensington–Gloucester Road is used only by the Circle Line), and the line was then extended in stages, reaching Tower Hill in 1882. The last few decades of the 19th century also saw extensions to Ealing Broadway, Kensington Olympia, Hounslow, Richmond and Wimbledon, followed by Upminster in 1902, Rayners Lane via Park Royal in 1910, and in 1926 the Wimbledon–High Street Kensington service was extended to Edgware Road (although this section had already been open as part of the Inner Circle since 1871 – see Circle Line).

For many years the Hounslow West branch (since extended to Heathrow Airport) was served both by District and Piccadilly Line trains, until the District Line ceased to operate to Hounslow in 1964. The Acton Town–Rayners Lane via Park Royal line was served by the District Railway for its first 23 years of existence before being taken over by the Piccadilly Line in 1933.

The only part of the District Line now closed, and not served by any other lines, is the Acton Town–South Acton shuttle, opened in 1905 as a through service between South Acton and Hounslow Barracks, reduced to a self-contained shuttle service in 1932 and withdrawn in 1959. Another unusual service operated by the District Line in the early part of the 20th century was the seasonal Ealing–Southend excursion trains which ran from 1910 to 1939. These were hauled by electric locos as far as Barking and then by steam.

The easternmost section between Bow Road and Upminster is above ground, as are the three western branches between Gloucester Road and Ealing Broadway, Richmond and Wimbledon. The central section between Gloucester Road and Bow Road, and the Edgware Road branch between Earl's Court and Edgware Road, are in tunnel, but there are gaps in the tunnels at South Kensington, Sloane Square, High Street Kensington, Notting Hill Gate, Edgware Road and Whitechapel stations.

SERVICES

Service patterns have altered little since the withdrawal of District Line services from the Hounslow branch. The District Line can be said to be broadly divided into two distinct parts: the District Main Line via Victoria, served by trains running between Ealing Broadway, Richmond or Wimbledon in the west and Barking, Dagenham East or Upminster in the east (with some trains starting and terminating at Tower Hill), and the Wimbledon–Edgware Road section. Wimbledon is the only one of the three western termini to be served by both Edgware Road and Main Line trains. The High Street Kensington–Olympia shuttle currently operates only at weekends and Bank Holidays, and on weekdays during major events at the Olympia exhibition centre (a limited early morning and evening service operates to and from Olympia on weekdays at other times). The hub of the District Line can be said to be Earl's Court. On Mondays–Saturdays the first

▲ S8 Stock 21039/040 arrives at Chorleywood with a Metropolitan Line service for Aldgate on 9 September 2017. **Jamie Squibbs**

▼ Clearly showing the different dimensions between Sub Surface and Tube stock, on 8 November 2018 96103 leads a Southbound Jubilee Line train to North Greenwich at Kilburn as S Stock 21116/115 overtakes with a service to Aldgate. **Jamie Squibbs**

trains begin their journeys at around 04.50 and the last trains reach their destinations between 00.45 and 01.40. Sunday services run from around 06.00 to midnight. The basic off peak service consists of:

- Six trains per hour Ealing Broadway–Upminster;
- Six trains per hour Richmond–Upminster;
- Six trains per hour Wimbledon–Edgware Road;
- Six trains per hour Wimbledon–Tower Hill (three of which continue to Barking);
- Three trains per hour High Street Kensington–Olympia at weekends only.

76 trains are required for the peak District Line service.

1.8.3. HAMMERSMITH & CITY LINE

Route: Hammersmith–Barking. **Route Length**: 16 miles.
No. of trains: 133 x 7-car (931 cars)*. **No. of stations**: 29. **First section opened**: 1863.
* Fleet shared with the Circle and District Lines.

The Hammersmith & City Line runs from Hammersmith in west London via Paddington and the top part of the Circle Line to Liverpool Street, and then along the District Line tracks to Barking.

The western section of the Hammersmith & City Line, between Paddington and Hammersmith, was opened in 1864 by the Hammersmith & City Railway, a subsidiary of the Great Western Railway. It is thus the second oldest section of the London Underground, having opened only a year after the Metropolitan Railway's initial stretch from Paddington to Farringdon Street, which does now also form part of the Hammersmith & City Lines route. For its first year of operation, the Hammersmith branch ran as a broad gauge line with through services to Farringdon Street on the MR. In 1865 the Hammersmith branch became standard (4 ft 8½ in/1435 mm) gauge and a broad gauge service between Paddington and Addison Road (now Kensington Olympia) started. The Addison Road service used a spur onto the West London Line from Latimer Road, closed in 1940.

The short section from Liverpool Street to Aldgate East is the only section of the line used exclusively by H&C trains. Apart from this, Edgware Road–Hammersmith was also not used by any other lines until the extension of Circle Line services to Hammersmith in 2009. All other sections served by the H&C have always been shared with other lines: Edgware Road–Liverpool Street is also used by the Circle Line (plus the Metropolitan Line east of Baker Street) and Aldgate East–Barking is shared with the District Line.

The Hammersmith & City Line formed part of the Metropolitan Line until 1990 when it received its own separate identity and pink colour on the LU maps.

All of the Hammersmith branch is above ground apart from the Subway Tunnel, which takes the line under the Great Western Main Line between Royal Oak and Westbourne Park. At the west end much of the route is elevated on viaducts, giving good views of the west London cityscape.

SERVICES

Historically, H&C trains have only continued beyond Whitechapel in peak hours, but today almost all trains run along the entire length of the line between Hammersmith and Barking.

The basic off-peak service pattern sees six trains per hour between Hammersmith and Barking, running at intervals of between 9 and 11 minutes. Combined with the Circle Line this gives 12 trains per hour between Hammersmith and Liverpool Street via Edgware Road and Baker Street. On Mondays–Saturdays the first trains leave Hammersmith as early as 04.37 and Barking at 05.01, and the last trains terminate at Hammersmith at 01.10 and Barking at 01.19. On Sundays the first trains leave Barking at 06.26 and Hammersmith at 06.24 (preceded by the first Circle Line train from Hammersmith at 06.21) and the last trains leave Barking at 23.15 (followed by two more trains starting from Plaistow at 23.38 and 23.53) and Hammersmith at 23.55. 33 trains are required for the peak Circle and Hammersmith & City Line services.

1.8.4. METROPOLITAN LINE

Route: Uxbridge/Amersham/Chesham/Watford–Aldgate. **Route Length:** 41½ miles.
No. of trains: 59 x 8-car (472 cars). **No. of stations:** 34. **First section opened:** 1863.

The Metropolitan Line operates longer distance services from the far suburbs of Amersham, Chesham and Watford, and also Uxbridge, into London, via Baker Street and King's Cross, to Aldgate (for the City).

As its name suggests, the Metropolitan Line has its origins in its namesake railway company, the Metropolitan Railway. Having opened its initial stretch of line between Paddington and Farringdon in 1863, the first stretch of today's Metropolitan Line north of Baker Street was completed in 1868 as far as Swiss Cottage. This was followed by Swiss Cottage–West Hampstead 11 years later, and the line was then gradually extended northwards until Rickmansworth was reached in 1887, Chesham in 1889 and Amersham (along with the northernmost extensions to Aylesbury and Verney Junction) in 1892. The Harrow-on-the-Hill–Uxbridge branch opened in 1904 and Moor Park–Watford in 1925. Also served by the MR was the Brill branch but this never had a direct service to central London. From the 1910s onwards the entire area north of Baker Street served by the MR came to be known as Metroland.

Services to Verney Junction were withdrawn in 1936, and in 1961 the Metropolitan Line ceased to serve Aylesbury which has since been served only by trains from Marylebone on the remaining southernmost section of the erstwhile Great Central Main Line. These trains, operated by Chiltern Railways, have interchange with the Metropolitan Line at Amersham and Harrow-on-the-Hill, with LU and Chiltern sharing the same platforms at the intermediate stations at Chalfont & Latimer, Chorleywood and Rickmansworth. All of these stations are managed by London Underground, however.

As mentioned above, the Baker Street–Hammersmith line was also classed as part of the Metropolitan Line until 1990 when it was redesignated as the Hammersmith & City Line. Since then the Metropolitan Line has consisted entirely of the former MR lines north from Baker Street plus the top right-hand corner of the circle between Baker Street and Aldgate which is shared with Circle Line trains (and with the H&C as far as Liverpool Street).

The entire Metropolitan Line north of Finchley Road is above ground. Finchley Road–Baker Street–Aldgate is in tunnel apart from two short gaps at Farringdon and Barbican stations. Amersham is the highest station on the LU network at 147 metres above sea level, and the Metropolitan Line also has the longest distance between two LU stations: 6.3 km (4 miles) between Chesham and Chalfont & Latimer.

SERVICES

Historically, the Metropolitan Line has been a largely self-contained operation, with trains only continuing east of Baker Street in peak hours. Since 1990 trains have continued beyond Baker Street to Aldgate throughout the day, supplementing the Circle and Hammersmith & City Line trains on the northern part of the Circle Line and giving up to 30 trains per hour on the Baker Street–Liverpool Street section at peak times.

At present, off-peak Watford trains start and terminate at Baker Street, with most trains for Uxbridge and all those for Chesham and Amersham continuing to and from Aldgate. During peak hours the service pattern is more complex, consisting of a mixture of through trains to Aldgate and Baker Street terminators to and from all destinations.

Since December 2010, the Chesham branch has seen regular all-day through trains to and from Aldgate, having previously been served only by a shuttle from Chalfont & Latimer. This had the effect of halving the off-peak service frequency at Amersham from four to two trains per hour (although these are complemented by Chiltern Railways' two trains per hour London Marylebone–Aylesbury service), with the other two trains serving Chesham. Watford is served by four trains per hour and Uxbridge eight.

Another major change to the long-established service pattern was made in December 2011, with all off-peak trains calling at all stations between Finchley Road and Moor Park. Until that time, Watford trains ran semi-fast and Amersham trains were designated as "fast", but now only certain peak hour trains (a.m. peak southbound, p.m. peak northbound) pass non-stop through some stations. Trains classed as semi-fast are non-stop between Wembley Park and Harrow-

▲ One of the most idyllic scenes on London Underground – a train of S8 stock crosses the Grand Union Canal at Croxley with a service from Watford to Aldgate on 21 April 2016. This section of line was due to close as part of the Croxley link scheme, but plans for this have been shelved. **David Palmer**

on-the-Hill, with "fast" trains calling only at Harrow-on-the-Hill between Wembley Park and Moor Park. Southbound fast and semi-fast trains in the morning peak do not call at Wembley Park.

An extension of the Watford branch to Watford Junction was planned, but at the time of going to press this project, known as the Croxley Rail Link, had effectively been shelved because of TfL funding constraints. The scheme would have involved the closure of the existing Metropolitan Line station at Watford and construction of a link between the LU branch and the DC Lines at Watford High Street with two new stations at Cassiobridge and Watford Vicarage Road. Instead the Mayor of Watford has called for alternative solutions such as a light rail or bus rapid transit based scheme to be examined. An extension of the Metropolitan Line to Watford town centre had also been considered on a number of previous occasions, firstly by the Metropolitan Railway in 1927 and then by the London Transport Executive in 1948. The second of these schemes would have involved linking the Metropolitan Line Watford branch to the now closed Watford Junction–Croxley Green branch. A similar scheme was proposed by London Underground in 1994, which eventually evolved into the Croxley Rail Link project.

SUB-SURFACE LINE ROLLING STOCK AND DEPOTS

All Sub-Surface Line trains are formed of S7 or S8 Stock built by Bombardier, Derby under the "Movia" marketing name. These trains are air-conditioned and all axles are powered. It is possible to walk through an entire train of S Stock, unlike other units which are formed of individual cars separated by emergency doors.

On the Metropolitan Line the S8s replaced the 1960s-built A Stock by 2012. The S8 Stock differs from the Circle, District and Hammersmith & City Line S7 Stock in having a mixture of facing and longitudinal seating. The S7 Stock replaced the 1960s/70s C Stock, the last of which was withdrawn in 2014 and also the D78 Stock. The last D78 stock was withdrawn from traffic in April 2017. More than 200 D78 Stock vehicles were acquired by Vivarail and some are being rebuilt as main line diesel or battery-diesel Class 230 units. These are listed in other Platform 5 publications.

8-car (S8) trains are used on the Metropolitan Line only, 7-car (S7) trains are used on the Circle, District and Hammersmith & City lines. All S7 stock operates in a common user pool. All S Stock is in permanent formations and does not uncouple, including when lifted for depot examinations.

There were originally 58 S8 trains built. Two trains of S7 Stock were formed as 8-car sets (known as S7+1 Stock, see note below). 49 trains are required for the Metropolitan Line peak service. The main depot for these is Neasden; trains also stable at Rickmansworth, Watford, Uxbridge, and Wembley Park.

In total there are 132 S7 trains. They entered service first on the Hammersmith & City Line in 2012, followed by the Circle Line and finally the District Line. The Circle and Hammersmith & City Lines are jointly managed, and 33 of the S7s are required for these two lines peak service. For maintenance purposes the S7s are based at the District Line depots at Ealing Common and Upminster, and can also receive attention at the Metropolitan Line depot at Neasden. For the Hammersmith & City and Circle the main depot for stabling is at Hammersmith. Trains also stable at Ealing Common, Triangle Sidings (Gloucester Road), Wembley Park, Neasden, Edgware Road, Moorgate, Aldgate, Barking, and Upminster.

76 trains are required for the District Line peak service (unusually more are required in the evening peak compared to the morning peak (75 sets)). The main depots are at Ealing Common and Upminster; trains also stable at Richmond, Parsons Green, Lillie Bridge, Triangle Sidings, High Street Kensington, and Barking.

A note on the S7+1 sets: The original order for S Stock was for 133 7-car trains and 58 8-car trains (1395 vehicles). Later an additional train was added to the order to cover for an additional unit required for the now shelved Croxley extension. The additional train was built as a standard S7 unit (21567/568), tagged onto the end of the S7 build. However, in order to create two additional trains for the Metropolitan Line (one for the Croxley link and another to provide an extra set whilst modifications were made to the fleet) it was decided to augment two S7s to 8-cars (called S7+1). These were units 21323/324 and 21327/328, which were formed using the M2 vehicles from sets 21383/384 and 21385/386. New replacement vehicles were built for 21383/384 and 21385/386 (numbered 23384 and 23386). A quirk of these formations was that 21323/324 and 21327/328 (the two S7+1s) had two Sandite cars. This also meant that two trains used only on the Metropolitan Line had all longitudinal seating, unlike the S8 Stock. These two sets were later reformed as standard S7 sets but in June 2019 21323/324 was reformed again as an S7+1 for the Metropolitan Line, this time using 25386 as an extra M2 car. At the time of writing 25384 remains stored out of use at Derby.

S STOCK — SUB-SURFACE LINES

Formations: S8: DM–M1–M2–MS–MS–M2–M1–DM.
S7: DM–M1–MS–MS–M2–M1–DM. S7+1: DM–M1–M2–MS–MS–M2–M1–DM.
Construction: Bolted and welded aluminium extrusions.
Traction Motors: 4 x AC fully suspended Bombardier MJB20093 of 65 kW (87 hp) per motor car.
Wheel Arrangement: Bo-Bo + Bo-Bo + Bo-Bo + Bo-Bo + Bo-Bo (+ Bo-Bo) + Bo-Bo + Bo-Bo.
Braking: Analogue rheostatic and regenerative.
Dimensions: 17.44/15.43 x 2.92 m. **Height**: 3.682 m.
Bogies: Flexible frame (Bombardier).
Gangways: Wide walk-through within unit; end emergency exit.
Couplers: LU automatic wedgelock with pneumatic connections. Two S Stock units may couple and multiple for rescue and tractor purposes.
Control System: IGBT.
Doors: Sliding plug.
Maximum Speed: S8: 62 mph; S7: 45 mph.
Train Protection: Thales Seltrac CBTC.
Seating Layout: S7: 212 seats longitudinal, plus 44 tip-ups;
S7+1: 242 seats longitudinal, plus 50 tip-ups;
S8: 256 seats (some 2+2 facing one side, opposite longitudinal/tip-ups), plus 50 tip-ups.

Cars in the 25xxx number series are fitted with de-icing equipment.

S8 – 8-car units

DM. Bombardier 2008–12. 34 seats + 6 tip-ups. 33.3 tonnes.
M1. Bombardier 2008–12. 32 seats + 6 tip-ups. 30.6 tonnes.
M2. Bombardier 2008–12. 32 seats + 6 tip-ups. 27.5 tonnes.
MS. Bombardier 2008–12. 30 seats + 7 tip-ups. 29.2 tonnes.

DM	*M1*	*M2*	*MS*	*MS*	*M2* †	*M1*	*DM*
21001	22001	23001	24001	24002	25002	22002	21002
21003	22003	23003	24003	24004	25004	22004	21004
21005	22005	23005	24005	24006	25006	22006	21006
21007	22007	23007	24007	24008	25008	22008	21008
21009	22009	23009	24009	24010	25010	22010	21010
21011	22011	23011	24011	24012	25012	22012	21012
21013	22013	23013	24013	24014	25014	22014	21014
21015	22015	23015	24015	24016	25016	22016	21016
21017	22017	23017	24017	24018	25018	22018	21018
21019	22019	23019	24019	24020	25020	22020	21020
21021	22021	23021	24021	24022	25022	22022	21022
21023	22023	23023	24023	24024	25024	22024	21024
21025	22025	23025	24025	24026	25026	22026	21026
21027	22027	23027	24027	24028	25028	22028	21028
21029	22029	23029	24029	24030	25030	22030	21030
21031	22031	23031	24031	24032	25032	22032	21032
21033	22033	23033	24033	24034	25034	22034	21034
21035	22035	23035	24035	24036	25036	22036	21036
21037	22037	23037	24037	24038	25038	22038	21038
21039	22039	23039	24039	24040	25040	22040	21040
21041	22041	23041	24041	24042	25042	22042	21042
21043	22043	23043	24043	24044	25044	22044	21044
21045	22045	23045	24045	24046	25046	22046	21046

21047	22047	23047	24047	24048	25048	22048	21048
21049	22049	23049	24049	24050	25050	22050	21050
21051	22051	23051	24051	24052	25052	22052	21052
21053	22053	23053	24053	24054	25054	22054	21054
21055	22055	23055	24055	24056	25056	22056	21056
21057	22057	23057	24057	24058	23058	22058	21058
21059	22059	23059	24059	24060	23060	22060	21060
21061	22061	23061	24061	24062	23062	22062	21062
21063	22063	23063	24063	24064	23064	22064	21064
21065	22065	23065	24065	24066	23066	22066	21066
21067	22067	23067	24067	24068	23068	22068	21068
21069	22069	23069	24069	24070	23070	22070	21070
21071	22071	23071	24071	24072	23072	22072	21072
21073	22073	23073	24073	24074	23074	22074	21074
21075	22075	23075	24075	24076	23076	22076	21076
21077	22077	23077	24077	24078	23078	22078	21078
21079	22079	23079	24079	24080	23080	22080	21080
21081	22081	23081	24081	24082	23082	22082	21082
21083	22083	23083	24083	24084	23084	22084	21084
21085	22085	23085	24085	24086	23086	22086	21086
21087	22087	23087	24087	24088	23088	22088	21088
21089	22089	23089	24089	24090	23090	22090	21090
21091	22091	23091	24091	24092	23092	22092	21092
21093	22093	23093	24093	24094	23094	22094	21094
21095	22095	23095	24095	24096	23096	22096	21096
21097	22097	23097	24097	24098	23098	22098	21098
21099	22099	23099	24099	24100	23100	22100	21100
21101	22101	23101	24101	24102	23102	22102	21102
21103	22103	23103	24103	24104	23104	22104	21104
21105	22105	23105	24105	24106	23106	22106	21106
21107	22107	23107	24107	24108	23108	22108	21108
21109	22109	23109	24109	24110	23110	22110	21110
21111	22111	23111	24111	24112	23112	22112	21112
21113	22113	23113	24113	24114	23114	22114	21114
21115	22115	23115	24115	24116	23116	22116	21116

† De-icing vehicles (25xxx) are designated M2D.

Name:

21100 Tim O' Toole CBE

S7 – 7-car units (or * S7+1 8-car units)

DM. Bombardier 2011–15. 32 seats + 6 tip ups. 33.3 tonnes.
M1. Bombardier 2011–15. 30 seats + 6 tip-ups. 30.6 tonnes.
M2. Bombardier 2011–15. 30 seats + 6 tip-ups. 27.5 tonnes.
MS. Bombardier 2011–15. 29 seats + 7 tip-ups. 29.2 tonnes.

DM	*M1*	*M2* †	*MS*	*MS*	*M2* †	*M1*	*DM*
21301	22301		24301	24302	25302	22302	21302
21303	22303		24303	24304	25304	22304	21304
21305	22305		24305	24306	25306	22306	21306
21307	22307		24307	24308	25308	22308	21308
21309	22309		24309	24310	25310	22310	21310
21311	22311		24311	24312	25312	22312	21312
21313	22313		24313	24314	25314	22314	21314
21315	22315		24315	24316	25316	22316	21316
21317	22317		24317	24318	25318	22318	21318
21319	22319		24319	24320	25320	22320	21320
21321	22321		24321	24322	25322	22322	21322
21323	22323	25386	24323	24324	25324	22324	21324*
21325	22325		24325	24326	25326	22326	21326
21327	22327		24327	24328	25328	22328	21328
21329	22329		24329	24330	25330	22330	21330
21331	22331		24331	24332	25332	22332	21332
21333	22333		24333	24334	25334	22334	21334

▲ S7 Stock 21370/369 arrives into South Kensington with a District Line service to Tower Hill on 16 March 2019. **Robert Pritchard**

▼ S7 Stock 21392/391 arrives at the original and superbly preserved Metropolitan Line Platforms 5 and 6 at Baker Street (that date from 1863) on 15 March 2019 with a service for Barking. **Robert Pritchard**

21335	22335	24335	24336	25336	22336	21336
21337	22337	24337	24338	25338	22338	21338
21339	22339	24339	24340	25340	22340	21340
21341	22341	24341	24342	25342	22342	21342
21343	22343	24343	24344	25344	22344	21344
21345	22345	24345	24346	25346	22346	21346
21347	22347	24347	24348	25348	22348	21348
21349	22349	24349	24350	25350	22350	21350
21351	22351	24351	24352	25352	22352	21352
21353	22353	24353	24354	25354	22354	21354
21355	22355	24355	24356	25356	22356	21356
21357	22357	24357	24358	25358	22358	21358
21359	22359	24359	24360	25360	22360	21360
21361	22361	24361	24362	25362	22362	21362
21363	22363	24363	24364	25364	22364	21364
21365	22365	24365	24366	25366	22366	21366
21367	22367	24367	24368	25368	22368	21368
21369	22369	24369	24370	25370	22370	21370
21371	22371	24371	24372	25372	22372	21372
21373	22373	24373	24374	25374	22374	21374
21375	22375	24375	24376	25376	22376	21376
21377	22377	24377	24378	25378	22378	21378
21379	22379	24379	24380	25380	22380	21380
21381	22381	24381	24382	25382	22382	21382
21383	22383	24383	24384	23384	22384	21384
21385	22385	24385	24386	23386	22386	21386
21387	22387	24387	24388	23388	22388	21388
21389	22389	24389	24390	23390	22390	21390
21391	22391	24391	24392	23392	22392	21392
21393	22393	24393	24394	23394	22394	21394
21395	22395	24395	24396	23396	22396	21396
21397	22397	24397	24398	23398	22398	21398
21399	22399	24399	24400	23400	22400	21400
21401	22401	24401	24402	23402	22402	21402
21403	22403	24403	24404	23404	22404	21404
21405	22405	24405	24406	23406	22406	21406
21407	22407	24407	24408	23408	22408	21408
21409	22409	24409	24410	23410	22410	21410
21411	22411	24411	24412	23412	22412	21412
21413	22413	24413	24414	23414	22414	21414
21415	22415	24415	24416	23416	22416	21416
21417	22417	24417	24418	23418	22418	21418
21419	22419	24419	24420	23420	22420	21420
21421	22421	24421	24422	23422	22422	21422
21423	22423	24423	24424	23424	22424	21424
21425	22425	24425	24426	23426	22426	21426
21427	22427	24427	24428	23428	22428	21428
21429	22429	24429	24430	23430	22430	21430
21431	22431	24431	24432	23432	22432	21432
21433	22433	24433	24434	23434	22434	21434
21435	22435	24435	24436	23436	22436	21436
21437	22437	24437	24438	23438	22438	21438
21439	22439	24439	24440	23440	22440	21440
21441	22441	24441	24442	23442	22442	21442
21443	22443	24443	24444	23444	22444	21444
21445	22445	24445	24446	23446	22446	21446
21447	22447	24447	24448	23448	22448	21448
21449	22449	24449	24450	23450	22450	21450
21451	22451	24451	24452	23452	22452	21452
21453	22453	24453	24454	23454	22454	21454
21455	22455	24455	24456	23456	22456	21456
21457	22457	24457	24458	23458	22458	21458
21459	22459	24459	24460	23460	22460	21460
21461	22461	24461	24462	23462	22462	21462

21463	22463	24463	24464	23464	22464	21464
21465	22465	24465	24466	23466	22466	21466
21467	22467	24467	24468	23468	22468	21468
21469	22469	24469	24470	23470	22470	21470
21471	22471	24471	24472	23472	22472	21472
21473	22473	24473	24474	23474	22474	21474
21475	22475	24475	24476	23476	22476	21476
21477	22477	24477	24478	23478	22478	21478
21479	22479	24479	24480	23480	22480	21480
21481	22481	24481	24482	23482	22482	21482
21483	22483	24483	24484	23484	22484	21484
21485	22485	24485	24486	23486	22486	21486
21487	22487	24487	24488	23488	22488	21488
21489	22489	24489	24490	23490	22490	21490
21491	22491	24491	24492	23492	22492	21492
21493	22493	24493	24494	23494	22494	21494
21495	22495	24495	24496	23496	22496	21496
21497	22497	24497	24498	23498	22498	21498
21499	22499	24499	24500	23500	22500	21500
21501	22501	24501	24502	23502	22502	21502
21503	22503	24503	24504	23504	22504	21504
21505	22505	24505	24506	23506	22506	21506
21507	22507	24507	24508	23508	22508	21508
21509	22509	24509	24510	23510	22510	21510
21511	22511	24511	24512	23512	22512	21512
21513	22513	24513	24514	23514	22514	21514
21515	22515	24515	24516	23516	22516	21516
21517	22517	24517	24518	23518	22518	21518
21519	22519	24519	24520	23520	22520	21520
21521	22521	24521	24522	23522	22522	21522
21523	22523	24523	24524	23524	22524	21524
21525	22525	24525	24526	23526	22526	21526
21527	22527	24527	24528	23528	22528	21528
21529	22529	24529	24530	23530	22530	21530
21531	22531	24531	24532	23532	22532	21532
21533	22533	24533	24534	23534	22534	21534
21535	22535	24535	24536	23536	22536	21536
21537	22537	24537	24538	23538	22538	21538
21539	22539	24539	24540	23540	22540	21540
21541	22541	24541	24542	23542	22542	21542
21543	22543	24543	24544	23544	22544	21544
21545	22545	24545	24546	23546	22546	21546
21547	22547	24547	24548	23548	22548	21548
21549	22549	24549	24550	23550	22550	21550
21551	22551	24551	24552	23552	22552	21552
21553	22553	24553	24554	23554	22554	21554
21555	22555	24555	24556	23556	22556	21556
21557	22557	24557	24558	23558	22558	21558
21559	22559	24559	24560	23560	22560	21560
21561	22561	24561	24562	23562	22562	21562
21563	22563	24563	24564	23564	22564	21564
21565	22565	24565	24566	23566	22566	21566
21567	22567	24567	24568	23568	22568	21568

† De-icing vehicles (25xxx) are designated M2D.

Spare 25384 (S) Bombardier Derby

▲ S7 Stock 21480/479 arrives at Farringdon with a service to Hammersmith on 14 February 2019. **Robert Pritchard**

▼ S8 Stock interior, showing MS 24002 of the first-built set, 21001/002. **Robert Pritchard**

1.9. NON-PASSENGER STOCK

This section contains details of former passenger carrying stock converted to departmental use and also engineering locomotives and purpose-built On-Track Machines.

Engineering trains operate on LU during the very limited hours when the network is not opened to passengers. Most trains are top-and-tailed by two of the battery electric locomotives. These normally travel from Ruislip depot to the worksite under electric power, then operate in battery power within a possession.

1.9.1. ENGINEERS' STOCK

TRACK RECORDING TRAIN ex-1960 & ex-1973 STOCK

3-car Track Recording Unit TRV (Track Recording Vehicle) formed of two former 1960 Stock motor cars either side of one former 1973 Stock trailer. Converted 1987 to monitor track condition.

Formation: 3-car; DM–T–DM.
Traction Motors: 4 x DC nose suspended GEC LT113 of 49 kW (60 hp) per motor car.
Wheel Arrangement: Bo-Bo + 2-2 + Bo-Bo.
Bogies: Plate frame.
Control System: Resistor camshaft.
Maximum Speed: 60 mph.

L132 ex-3901 | TRC666 ex-514 | L133 ex-3905

RAIL ADHESION TRAINS ex-1962 and ex-1956 STOCK

Two Central Line Sandite units, one 8-car and one 5-car, adapted from 4-car 1962 Tube Stock plus a car from similar 1956 Stock. The 1962 motor cars were built by Metropolitan-Cammell, the trailers by BR at Derby Works. The 8-car train is based at Hainault Depot for working the east end of the Central Line and the 5-car unit at Rusilip Depot for operating the west end.

Formation: 8-car; DM–T–NDM–DM–DM–NDM–T–DM; 5-car; DM–NDM–T–NDM–DM.
Construction: Steel body on steel underframe.
Traction Motors: 2 x DC nose-suspended Brush LT112 of 60 kW (80 hp) per motor car.
Wheel Arrangement: 8-car; 1A-A1 + 2-2 + 1A-A1 + 1A-A1 + 1A-A1 + 1A-A1 + 2-2 + 1A-A1; 5-car; 1A-A1 + 1A-A1 + 2-2 + 1A-A1 + 1A-A1.
Bogies: Plate frame. **Control System**: Resistor camshaft.

* Converted from 1956 Stock

1406	2682	9125*	1681	1682	9577	2406	1407	8-car Rail Adhesion Train
1570	9691	2440	9441	1441				5-car Rail Adhesion Train

ASSET INSPECTION TRAIN ex-1967 and ex-1972 STOCK

These vehicles have been undergoing a protracted conversion into a new 3-car Asset Inspection Train to supplement, or eventually replace, the Track Recording Train, but at the time of writing this work has stopped and the vehicles are stored at Northfields Depot.

Formation: 3-car; DM–T–DM.
Construction: Aluminium body on steel underframe.
Traction Motors: 4 x DC nose-suspended Brush LT115 of 53 kW (71 hp) per motor car.
Wheel Arrangement: Bo-Bo + 2-2 + Bo-Bo.
Braking: Electro-pneumatic and rheostatic.
Dimensions: 16.09/15.98 x 2.64 m.
Bogies: Plate frame.
Control System: Resistor camshaft.

* Converted from 1967 Stock (all others 1972 Stock).

3213 4213 3179* | 3079* 4313 3313

FILMING UNIT ex-1972 STOCK

4-car unit retained for movie filming purposes at Aldwych.

Formation: DM–T–T–DM.
Construction: Aluminium Body on steel underframe.
Traction Motors: 4 x DC nose-suspended Brush LT115 of 53 kW (71 hp) per motor car.
Wheel Arrangement: Bo-Bo + 2-2 + 2-2 + Bo-Bo.
Braking: Electro-pneumatic and rheostatic.
Bogies: Plate frame.
Couplers: LU automatic wedgelock.
Doors: Sliding.
Train Protection: Tripcock.
Dimensions: 16.09/15.98 x 2.64 m.
Gangways: Emergency end doors only.
Control System: Resistor camshaft.
Maximum Speed: 45 mph.

3229 | 4229 | 4329 | 3329

RAIL ADHESION TRAINS ex-D78 STOCK

Two 4-car former District Line D78 units converted and augmented to 5-car rail adhesion trains (or "RATS") for Sandite dispersal, replacing the previous trains of A60 Stock used on the Sub Surface Lines. Based at Neasden Depot, along with three spare vehicles.

Formation: 5-car; DM–UNDM–T–UNDM–DM.
Construction: Aluminium body on aluminium underframe.
Traction Motors: 4 x DC nose-suspended Brush LT118 of 49 kW (60 hp) per motor car.
Wheel Arrangement: Bo-Bo + 2-2 + Bo-Bo + 2-2 + Bo-Bo.
Dimensions: 18.37/18.12 x 2.85 m.
Bogies: Box frame.
Control System: Resistor camshaft.

7010 8123 17010 8010 7123 | 7040 8107 17040 8040 7107

Spares: 7526 17526 7527

▲ Track Recording Train L132+TRC666+L133 at North Acton on 30 April 2019. **Alisdair Anderson**

1.9.2. STORED FORMER LU STOCK

450 D78 Stock vehicles were built between 1980 and 1983, forming 75 6-car trains for the District Line. More than 220 former D78 Stock vehicles were purchased by Vivarail with a view to converting them to diesel or battery-diesel trains for the national railway network. Some have already been converted to Class 230 "D-Trains", whilst the other vehicles are stored at the Quinton Rail Technology Centre site at Long Marston in Warwickshire.

The following vehicles have so far been converted to Class 230s:

Prototype Units

230 001	300001 (7058)	300201 (17128)	300101 (7511)	To be exported to the USA (as a 2-car)
230 002	300002 (7122)		300102 (7067)	Battery prototype

Units operated by West Midlands Trains on the Bedford–Bletchley Line

230 003	300003 (7069)	300103 (7127)
230 004	300004 (7100)	300104 (7500)
230 005	300005 (7066)	300105 (7128)

Undergoing conversion for Transport for Wales

230 006	300006 (7098)	300206 (17066)	300106 (7510)
230 007	300007 (7103)	300207 (17063)	300107 (7529)
230 008	300008 (7120)	300208 (17050)	300108 (7065)
230 009	300009 (7055)	300209 (17084)	300109 (7523)
230 010	300010 (7090)	300210 (17071)	300110 (7017)

The following vehicles are stored at Long Marston:

DM cars (* New technology testbed vehicle; † Vehicle used for crash testing).

7000	7025	7050	7075	7095	7118	7516
7001	7026	7051	7076	7096	7119	7517
7002	7028	7052	7077	7097	7121	7518
7003	7029	7053	7078	7099	7124	7519
7004	7030	7054	7079	7101	7125	7520
7005	7031*	7056†	7080	7102	7126	7521
7006	7033	7057	7081	7104	7129	7522
7008	7034	7059	7082	7105	7501	7524
7009	7035	7060	7083	7106	7502	7525
7011	7038	7061	7084	7108	7503	7528
7013	7039	7062	7085	7109	7504	7530
7014	7041	7063	7086	7110	7505	7532
7015	7043	7064	7087	7111	7506	7533
7016	7044	7068	7088	7112	7507	7534
7019	7045	7070	7089	7113	7509	7535
7020	7046	7071	7091	7114	7512	7536
7021	7047	7072	7092	7115	7513	7537
7022	7048	7073	7093	7116	7514	7538
7023	7049	7074	7094	7117	7515	7539
7024						

T cars

17003	17035	17055	17069	17086	17100	17117
17011	17036	17056	17073	17087	17103	17120
17013	17043	17058	17076	17088	17104	17124
17015	17045	17059	17077	17089	17106	17127
17020	17047	17062	17078	17090	17110	17129
17021	17049	17064	17080	17091	17112	17510
17023	17052	17065	17082	17092	17113	17520
17025	17053	17067	17083	17096	17114	17522
17029	17054	17068	17085	17098	17116	17528

▲ Stored D78 Stock cars, owned by Vivarail, at the Quinton Rail Technology Centre site at Long Marston on 20 June 2019. DM cars 7077, 7039 and 7085 are nearest the camera. **Robert Pritchard**

▼ Schoma diesel-hydraulic locomotive 3 "CLAIRE" at Ruislip depot on 9 December 2015. **Keith Fender**

1.9.3. LOCOMOTIVES

SCHOMA DIESEL-HYDRAULIC or *CLAYTON BATTERY LOCOMOTIVES B or *Bo

Originally built to Tube gauge for construction of the Jubilee Line, then adsorbed into general engineers' stock. Built by Schoma in Germany as model CFL500VR diesel-hydraulic. Rebuilt from 2014 onwards to *battery electric by Clayton but at the time of writing all currently stored at Ruislip depot. Rebuilt machines have two underframe mounted traction motors driving the former hydraulic transmission cardan shafts. Rebuilds are designated model CB40 by Clayton and have 120 tonne load haulage capacity.

All 12 are named after secretaries of works managers on the Jubilee Line extension project.

Built: 1996; * rebuilt as battery electric.
Builder: Schoma; * rebuilt by Clayton.
Engine: As built – MTU 6V183TD 370 kW (495 hp); * battery.
Transmission: As built – diesel-hydraulic, Voith; * electric.
Continuous Rating: 208 kW.
Length: 8.50 m.
Weight: As built: 33.9 tonnes; * 40.0 tonnes.
Maximum Speed: As built: 31 mph; * 19 mph.

B diesel-hydraulic locomotives or * rebuilt as Bo battery electric locomotives

1 BRITTA LOTTA
2 * (S) NIKKI
3 CLAIRE
4 * (S) PAM
5 * (S) SOPHIE
6 * (S) DENISE
7 * (S) ANNEMARIE
8 * (S) EMMA
9 DEBORA
10 * (S) CLEMENTINE
11 * (S) JOAN
12 MELANIE
13 * (S) MICHELE
14 * (S) CAROL

BATTERY ELECTRIC LOCOMOTIVES Bo-Bo

These Tube gauge battery electric locomotives are based on a unique LT design first introduced in 1936. 52 were built to a similar basic design, but with each successive batch built up to 1974 incorporating progressive improvements (the 1985-built L62–L67 were unsuccessful and have since been scrapped). The 1936-built (L35–L43), 1951-built (L55–L61) and 1962-built (L76, later L33) batches are no longer in service. L35 is preserved (see section 2.4).

The underframes and bodies were newly built but traction motors and type Z bogies were recovered from passenger stock, except the 1973 batch which have new bogies built to the old design. Batches were built for specific construction projects, then absorbed into general use.

Batteries are recharged at depots; there is no charging off traction rails. Battery capacity has been progressively increased over time from 768 Ah to 1250 Ah. All locos have tripcocks, but many are fitted with line specific train protection systems. Equipped with 3 mph slow speed control.

Privatisation placed all battery locomotives into TransPlant ownership, part of TubeLines, itself now a TfL subsidiary. All battery locomotives are now based at Ruislip depot for both maintenance and operation. In 2010, a 15-year life extension and refurbishment of L24 was completed at Acton Works. L25 and L29 followed, and the remaining 26 were then completed by 2018. Refurbished locomotives are all fitted with new 168 cell Exide 1250 Ah batteries. Nominal traction rail voltage has been increased from 630 V to 750 V.

Current operations require 22 locomotives for weekends, based on 11 trains with one locomotive at each end. In addition to their L numbers, they carry a five digit number yy0xx, yy denoting year of build and xx the L serial number. Main line TOPS numbers were once allocated in the 970xx series, again with xx the L number; but they have long been deregistered. Further temporary numbers in the 977xx were allocated to some locomotives, also now deregistered, but all are shown here for completeness.

Built: p 1964; q 1969; r 1973.
Builder: p, q Metro-Cammell, Birmingham; r BREL Doncaster.
Battery: 1250 Ah lead acid.

▲ Battery-electric locomotives L26 and L32 top-and-tail a ballast train in Moorgate Platform 4 on 19 October 2017. **Jamie Squibbs**

▼ Cowans Boyd Self-propelled Twin-Jib Track Relayer TRM 627 is seen at Ruislip depot on 18 March 2009. **Kim Rennie**

Transmission: Electric or battery.
Traction Motors: 4 GEC WT54D of 112 kW continuous.
Maximum Tractive Effort: 17800 lbf (original build).
Continuous Rating: 600 hp (original build).
Length: 16.54 m (original build).
Weight: 53.8 tonnes (original build).
Wheel Diameter: 914 mm.
Maximum Speed: 30 mph.

p 1969 batch originally delivered for final LT steam replacement.
q 1964 batch originally delivered for Victoria Line construction.
r 1973 batch originally delivered for Fleet Line and Heathrow construction.
v Victoria Line ATP.

L15	pt	69015	97015, 97715
L16	pt	69016	97016
L17	pt	69017	97017
L18	pt	69018	97018
L19	pt	69019	97019
L20	qt	64020	97020, 97720
L21	qt	64021	97021
L22	q	64022	97022, 97722
L23	q	64023	97023
L24	q	64024	97024
L25	q	64025	97025
L26	q	64026	97026
L27	qv	64027	97027
L28	qv	64028	97028
L29	qv	64029	97029
L30	qv	64030	97030
L31	qv	64031	97031
L32	qv	64032	97032, 97732
L44	rt	73044	97044
L45	rt	73045	97045
L46	rt	73046	97046
L47	rt	73047	97047
L48	rt	73048	97048
L49	rt	73049	97049
L50	rt	73050	97050, 97750
L51	rt	73051	97051, 97751
L52	rt	73052	97052, 97752
L53	rt	73053	97053, 97754
L54	rt	73054	97054

1.9.4. ON-TRACK MACHINES

London Underground has a fleet of On-Track Machines operated through the Transplant business unit, all of which are based at Ruislip depot. These are used for maintaining, renewing and enhancing the network's infrastructure. Most machines are built to Tube profile, however the newer Tampers TMM774–TMM776 are built to Sub-Surface Line profile.

In addition other machines, particularly Rail Grinding Trains, are provided by specialist contractors for use on the network, but these machines are not listed here.

CRANES

C623	Cowans Sheldon 7.5 tonne Light Duty Diesel Hydraulic	Built: 1982
C624	Cowans Sheldon 7.5 tonne Light Duty Diesel Hydraulic	Built: 1984
C625	Cowans Sheldon 7.5 tonne Light Duty Diesel Hydraulic	Built: 1984
C626	Cowans Sheldon 7.5 tonne Light Duty Diesel Hydraulic	Built: 1984

TWIN-JIB TRACK RELAYERS

TRM627	Cowans Boyd Self-propelled	Built: 1986
TRM628	Cowans Boyd Self-propelled	Built: 1993

TAMPERS

TMM771	Plasser & Theurer 07-16 Universal		Built: 1980
TMM772	Plasser & Theurer 07-16 Universal		Built: 1980
TMM773	Plasser & Theurer 07-16 Universal	"ALAN JENKINS"	Built: 1980
TMM774	Plasser & Theurer 08-275/4 ZW		Built: 2007
TMM775	Matisa B45UE	EVN 99 70 9128 003-9	Built: 2015
TMM776	Matisa B45UE	EVN 99 70 9128 004-7	Built: 2015

2. PRESERVED UNDERGROUND STOCK

With a history dating back to 1863 it is not surprising that a number of former London Underground locomotives and trains are preserved at various museums and other private sites around the country.

The London Transport Museum's main site at Covent Garden in central London houses a small number of preserved Underground vehicles, trams, buses and trolleybuses along with a collection of artefacts from London's transport system through the ages. Most of the museum's preserved vehicles reside at the London Transport Museum Depot at Acton which typically has three open weekends per year, usually in April, July and September. Guided tours of the depot are also available on a number of other dates throughout the year. Details of events and attractions at both sites can be found at www.ltmuseum.co.uk.

There are also a small number of preserved London Transport vehicles at various other sites such as the Mangapps Railway Museum, Essex and the Buckinghamshire Railway Centre, Quainton Road. A few vehicles have even been converted for other uses, such as the D78 Stock car used as a library at Coopers Lane Primary School in Grove Park, South London and the 1983 Tube Stock cars in use as work rooms for artists and other creative practitioners at Village Underground on a part of the disused Broad Street viaduct in Shoreditch. A number of other vehicles are located at private sites.

Some Underground stock is still used in passenger service – the most famous are the six 2-car 1938 Stock trains used by South Western Railway on the Isle of Wight railway (detailed in other Platform 5 publications). Two 1959 Stock cars are also used as hauled stock in another unlikely location – the Alderney Railway in the Channel Islands.

Most preserved London Transport vehicles are on show as static exhibits only. However, a few of them are still in working order and are used on occasional special workings on the London Underground network, such as Metropolitan Railway steam locomotive No. 1, electric locomotive No. 12 "SARAH SIDDONS", the 4-car train of 1938 Tube Stock based at the LT Museum Depot at Acton and rake of vintage coaching stock used occasionally on the Sub Surface Lines. The last opportunity to travel on the loco-hauled stock into central London was in June 2019, as it will not be able to operate once the new communications based train control signalling system and automatic train operation (SelTrac CBTC) goes live. However, it is still planned to operate occasional steam specials on the outer reaches of the Metropolitan Line between Harrow-on-the-Hill and Amersham, where the fast lines are shared with Chiltern Railways trains and retain conventional signalling. The roll out of CBTC will however affect depot access, and it is currently planned to cease steam operations in 2021 when Metropolitan No. 1 is due an overhaul.

Other steam locomotives may also be used, and an ex-British Railways 4-car "4 TC" set is also based at London Underground's Ruislip depot for use as hauled stock (for details see the Platform 5 book "Preserved Locomotives of British Railways").

All London Underground EMU vehicles have an open saloon layout (usually including a mixture of facing and longitudinal seats) except where shown otherwise. Interior layouts are given for each type of loco-hauled passenger-carrying vehicle.

For a detailed list of locations see Appendix I.

2.1 METROPOLITAN RAILWAY STEAM LOCOMOTIVES

The Metropolitan Railway was a constituent of the present day London Underground Ltd, and is the only constituent company of LUL from which steam locomotives survive.

CLASS A — 4-4-0T

Built: 1866 by Beyer Peacock.
Boiler Pressure: 120 lb/sq. in.
Cylinders: 17¼" x 24" (O).
Valve Gear: Stephenson.
Wheel Diameters: 3′ 0″, 5′ 9″.
Weight: 42.15 tons.
Tractive Effort: 12 680 lbf.

23–L45	London Transport Museum	BP 710/1866

BRILL BRANCH — 4w WT

Built: 1872 by Aveling & Porter for use on Brill branch (Wotton Tramway).
Boiler Pressure:
Cylinders: 7¾" x 10".
Valve Gear: Stephenson.
Wheel Diameters: 3′ 0″.
Weight:
Tractive Effort:

–	Buckinghamshire Railway Centre	AP 807/1872

On loan from the London Transport Museum

CLASS E — 0-4-4T

Built: 1898 at Neasden. 7 built.
Boiler Pressure: 150 lb/sq. in.
Cylinders: 17¼" x 26" (I).
Valve Gear: Stephenson.
Wheel Diameters: 5′ 6″, 3′ 9½″.
Weight: 54.5 tons.
Tractive Effort: 15 420 lbf.

1–L44	Epping Ongar Railway	Neasden 3/1898

On loan from the Buckinghamshire Railway Centre.

2.2 LONDON UNDERGROUND DIESEL LOCOMOTIVES

London Underground Ltd has owned very few diesel locomotives, but two are now preserved.

UNCLASSIFIED — ROLLS–ROYCE — 0-6-0

Built: 1967–68. 3 built. Acquired 1971 from Thomas Hill, Kilnhurst.
Engine: Rolls–Royce C8FL of 242 kW (325 hp).
Transmission: Hydraulic.
Maximum Tractive Effort: 128 kN (28 800 lbf).
Wheel Diameter: 1067 mm.
Weight: 48 tonnes.

DL 82	Hardingham Station, Norfolk	RR 10272/1967
DL 83	Nene Valley Railway	RR 10271/1967

▲ Metropolitan Railway No. 1 runs round its train at North Weald on the Epping Ongar Railway on 24 February 2019 during a gala weekend. 31438 can be seen behind. **Brian Garvin**

▼ Metropolitan Railway Class A 4-4-0T No. 23 is seen on display at the London Transport Museum, Covent Garden on 6 April 2019. **Robert Pritchard**

2.3 LONDON UNDERGROUND ELECTRIC LOCOS

CITY & SOUTH LONDON RAILWAY Bo

This locomotive is thought to be the original C&SLR No. 13 through research documents. It also carried No. 1 at one time.
Built: 1890 by Beyer-Peacock to Mather & Platt design. 14 built.
System: 500 V DC third rail system.
Gauge: 762 mm.
Maximum Tractive Effort:
Weight: 10.3 tons.
Continuous Tractive Effort:
Wheel Diameter: 686 mm.
Maximum Speed: 25 mph.
Continuous Rating:

13–1–13 London Transport Museum BP 1890

On loan from the Science Museum Group

WATERLOO & CITY RAILWAY Bo

Built: 1898. Siemens design for operation for LSWR on the Waterloo & City.
System: 750 V DC third rail system.
Traction Motors: Two Siemens 45 kW (60 hp).
Maximum Speed:
Wheel Diameter: 3 ' 4".

DS75–75S National Railway Museum, Shildon SM 6/1898

METROPOLITAN RAILWAY Bo-Bo

These locomotives, numbered 1–20, were the second use of these numbers by the Metropolitan Railway. Original numbers were 1–10 (MC/BW 1906) and 11–20 (MC/BTH 1907). One from each batch (6 and 17) were drastic rebuilds retaining very little of the originals. Regular locomotive-hauled operation ceased in 1961.
Built: 1922–23. 20 built.
Traction Motors: Metropolitan Vickers MV339.
System: 630 V DC 4-rail system.
Continuous Rating: 895 kW (1200 hp).
Maximum Tractive Effort: 100 kN (22 600 lbf).
Weight: 62.5 tons.
Continuous Tractive Effort: 65 kN (14 720 lbf).
Wheel Diameter: 1105 mm.
Maximum Speed: 65 mph.

5	JOHN HAMPDEN	London Transport Museum	VL 1922
12	SARAH SIDDONS	London Underground, Ruislip Depot	VL 1922

LONDON TRANSPORT EXECUTIVE Bo-Bo

Built: Converted 1964 from two standard tube stock DM cars 3080+3109, built 1931 by Metropolitan-Cammell. Special purpose shunter for Acton Works internal use, created by cutting the DM cars in half and joining the two motor/cab ends together.
System: 630 V DC 4-rail system.
Traction Motors: 4 x GEC WT54 of 125 kW (167 hp).
Wheel Diameter: 915 mm.

L11 Cravens Heritage Trains, Epping 1964

LONDON PASSENGER TRANSPORT BOARD Bo-2-2-Bo

Built: Converted 1940 at Acton Works from cars 3954 + 3995, originally Central London Railway cars 223 (Metropolitan-Cammell) and 265 (BRCW), built 1903. Sleet (de-icing locomotive); Inner bogies are non-load-bearing but carry de-icing equipment.
System: 630 V DC 4-rail system.
Continuous Rating:
Traction Motors: 4 x DC BTH GE66 of 93 kW (125 hp).
Maximum Tractive Effort:
Weight: 41.45 tonnes.
Continuous Tractive Effort:
Wheel Diameter: 890 mm.
Maximum Speed:

ESL 107 London Transport Museum Depot, Acton 1939

▲ City & South London Railway locomotive No. 13 is seen at the London Transport Museum, Covent Garden on 6 April 2019. **Robert Pritchard**

▼ Waterloo & City Railway 75S on display at the National Railway Museum, Shildon on 24 October 2017. **Robert Pritchard**

▲ Metropolitan Railway-built Bo-Bo electric locomotive No. 12 "SARAH SIDDONS" is used to haul Battery Electric locomotives L27 and L24 from Ruislip depot to Acton Works on 22 May 2017, seen here at Ruislip. **Jamie Squibbs**

▼ London Transport Executive Bo-Bo electric shunting locomotive L11 is now in yellow livery and plinthed near Epping station, where it is seen on 14 April 2019 as a Central Line train led by 91209 arrives at Epping station. **Robert Pritchard**

▲ Bo-Bo Battery Electric locomotive L35 at the London Transport Museum Depot at Acton on 27 April 2019.
Alan Yearsley

▶ Metropolitan Railway 1904 Stock Trailer No. 4 is seen at the London Transport Museum Depot at Acton during an open day on 27 April 2019.
Alan Yearsley

2.4. BATTERY ELECTRIC LOCOMOTIVES

LONDON PASSENGER TRANSPORT BOARD Bo-Bo

Built: 1938 by GRCW but mounted on former LT EMU bogies and uses former LT EMU traction motors. Operates off either battery or rail supply. 9 built.
Battery: 924 amp-hour capacity; 160 lead acid cells supplying 320 V.
Traction Motors: 4 x Metropolitan-Vickers MV145AZ of 113 kW (150 hp).
Weight: 54.6 tons. **Maximum Tractive Effort:** 58 kN (17800 lbf).
Wheel Diameter: 940 mm.
Maximum Speed: 30 mph (rail supply), 15 mph (battery supply).

L35 London Transport Museum Depot, Acton GRCW 1938

2.5. LONDON UNDERGROUND EMU STOCK

System: 630 V DC 4-rail system.

1904 STOCK METROPOLITAN RAILWAY

Built: 1904 for Metropolitan Railway for use on the Baker Street–Uxbridge section. Subsequently saw use with MoD at Shoeburyness as a conference coach. Body internally stripped and initially preserved at Passmore Edwards Museum, North Woolwich (1985–97) where it suffered fire damage before transfer to LT Museum collection. Some doubt as to true identity.

T 17.57 x 2.7 m 21 tons 16 cwt 48 seats.

MR	*LPTB*	*MoD*			
4	9486	3021-RGP007021	T	London Transport Museum Depot, Acton	BM 1904

Q23 STOCK DISTRICT RAILWAY

Built: Rebuilt 1935–40 from G Class stock, built 1923 for the District Railway.
Traction Motors: 2 x BTH GE69 of 149 kW (200 hp).

DM 15.77 x 2.62 m 29.5 tons 44 seats.

DR	*LER*	*LPTB*	*LTE*			
644	238	4148	4248	DM	London Transport Museum	GRCW 1923
662	274	4184		DM	London Transport Museum Depot, Acton	GRCW 1923

662 is currently stored at London Underground, Acton Works.

T STOCK METROPOLITAN RAILWAY

Built: 1932 for the Metropolitan Railway. Converted 1961 for use as a sleet locomotive numbered ESL 118.
Traction Motors: 2 x GEC WT54 of 157 kW (210 hp).
Layout: Non-corridor compartments.

Type	*Length x width*	*Weight*	*Seats (compartments)*
DM	16.00 x 2.79 m	?? tons	50(5)

MR	*LPTB*	*LTE*			
249	2749	ESL 118B	DM	Buckinghamshire Railway Centre	BRCW 1932
258	2758	ESL 118A	DM	Buckinghamshire Railway Centre	BRCW 1932

▲ District Railway Q23 Stock DM 4248 is seen on display at the London Transport Museum, Covent Garden on 6 April 2019. **Robert Pritchard**

▼ Metropolitan Railway T Stock DM 258 (partially restored) and DM 249 at the Buckinghamshire Railway Centre at Quainton Road on 3 March 2019. **Stuart Hicks**

▲ London Electric Railway Standard Stock DML134 at the London Transport Museum Depot at Acton on 27 April 2019. **Alan Yearsley**

▼ The serviceable 1938 Stock train is seen forming the 15.53 Amersham–Ealing Common railtour at Alperton on 9 September 2018, formed 10012, 012256, 12048 and 11012. **Jamie Squibbs**

STANDARD STOCK — LONDON ELECTRIC RAILWAY

Built: 1923–35 for London Electric Railway († London Passenger Transport Board). Those shown * were converted 1967 for the British Railways Isle of Wight Line.
Traction Motors: 2 x GEC WT54 of 112 kW (150 hp).

DM	16.70 x 2.60 m	29.5 tons	30 seats.
DT	15.70 x 2.60 m	19.0 tons	44 seats.
T	15.17 x 2.60 m	17.0 tons	48 seats.

LER	*LPTB*	*LTE*	*BR*			
297–3327	3327			DM	London Transport Museum Depot, Acton	MC 1927
320–3370	3370	L134		DM	London Transport Museum Depot, Acton	MC 1927
	3693†	L131		DM	London Transport Museum Depot, Acton	MC 1934
1789–5279	5279*		27	DT	London Transport Museum Depot, Acton	MC 1925
846–7296	7296*		49	T	London Transport Museum Depot, Acton	CL 1923

Q35 STOCK — LONDON PASSENGER TRANSPORT BOARD

Built: Rebuilt 1951 as Control Trailer from N Class stock, built 1936 for use on the Metropolitan Line Hammersmith–Barking section.

T	15.58 x 2.72 m	tons	44 seats.

LPTB	*LTE*			
8063	08063	T	London Transport Museum Depot, Acton	MC 1936

CO/CP STOCK — LONDON PASSENGER TRANSPORT BOARD

Built: 1938–39. CO Stock for the Hammersmith & City Line; CP Stock for the Metropolitan Line.
Traction Motors: 2 x MV145AZ of 113 kW (152 bhp).

DM	15.58 x 2.72 m	tons	40 seats.
T	15.58 x 2.72 m	tons	44 seats.

LPTB	*LTE*				
13028	53028	"CO"	DM	Buckinghamshire Railway Centre	BRCW 1938
013063	013063	"CO"	T	Buckinghamshire Railway Centre	GRCW 1938
14233	54233	"CP"	DM	Buckinghamshire Railway Centre	GRCW 1939
14256	54256	"CP"	DM	*Location unknown*	BRCW 1939

14256 is stored at an unknown location in Essex.

Q38 STOCK — LONDON PASSENGER TRANSPORT BOARD

Built: 1938 for the Metropolitan Line.
Traction Motors: 2 x GEC WT45B of 113 kW (152 bhp).

DM	15.58 x 2.72 m	tons	40 seats.

LPTB	*LTE*			
4416	L126	DM	London Transport Museum Depot, Acton	GRCW 1938
4417	L127	DM	London Transport Museum Depot, Acton	GRCW 1938

4416 is currently stored at London Underground, Acton Works.

1938 STOCK — LONDON PASSENGER TRANSPORT BOARD

Built: 1938–40 for the Bakerloo, Northern and Piccadilly Lines.
Traction Motors: 2 x Crompton Parkinson/GEC/BTH LT100 of 125 kW (170 hp).

DM	15.93 x 2.69 m	27.40 tons	42 seats.
M	15.60 x 2.60 m	25.90 tons	40 seats.
T	15.60 x 2.60 m	20.70 tons	40 seats.

LPTB	*LTE*			
10012	10012	DM	London Transport Museum Depot, Acton	MC 1938
11178	11178–11012	DM	London Transport Museum Depot, Acton	MC 1938
11182	11182	DM	London Transport Museum	MC 1939
92048	92048–12048	M	London Transport Museum Depot, Acton	MC 1938
012229	012229–4927	T	Cravens Heritage Trains, Epping	BRCW 1938
012256	012256	T	London Transport Museum Depot, Acton	BRCW 1938

012229 is based at London Underground, Northfields Depot.

Six 2-car sets are still operated by South Western Railway on the Isle of Wight, details of these are given in other Platform 5 titles.

R38 STOCK — LONDON TRANSPORT EXECUTIVE

Built: 1938 as Q38 Stock (T). Rebuilt 1950 as R38 stock for the District Line (DM).
Traction Motors: 2 x DC LT 111 of 60 kW (80 hp).

DM 15.58 x 2.97 m 34.3 tons ?? seats.

22624	DM	Mangapps Railway Museum	GRCW 1938

R49 SURFACE STOCK — LONDON TRANSPORT EXECUTIVE

Built: 1951–53 for the District Line.
Traction Motors: 2 x LT 111 of 60 kW (80 hp).

DM 15.58 x 2.72 m tonnes 40 seats.

21147	DM	Men At Work, Hanwell	MC 1953
22679	DM	London Transport Museum Depot, Acton	MC 1952

1959 TUBE STOCK — LONDON TRANSPORT EXECUTIVE

Built: 1959–61 for the Piccadilly Line.
Traction Motors: 2 x GEC LT112 of 60 kW (80 hp).

DM 15.98 x 2.60 m 26.62 tonnes 42 seats.

1018	DM	Finmere Station, Newton Purcell, Oxfordshire	MC 1959
1030	DM	Mangapps Railway Museum	MC 1959
1044	DM	Alderney Railway	MC 1959
1045	DM	Alderney Railway	MC 1959
1085 "1031"	DM	Epping Ongar Railway	MC 1959
1304	DM	Finmere Station, Newton Purcell, Oxfordshire	MC 1961
1305	DM	Longstowe Station, near Bourn, Cambridgeshire	MC 1961
2018	T	Woodlands, Coppleridge, Motcombe, Dorset	MC 1959
2044	T	Finmere Station, Newton Purcell, Oxfordshire	MC 1959
2304	T	Finmere Station, Newton Purcell, Oxfordshire	MC 1959
9305	M	Finmere Station, Newton Purcell, Oxfordshire	MC 1959

▲ R49 Surface Stock DM 22679 at the London Transport Museum Depot at Acton on 27 April 2019. **Alan Yearsley**

▼ A60 Stock DM 5034 at the London Transport Museum Depot at Acton on 27 April 2019. **Alan Yearsley**

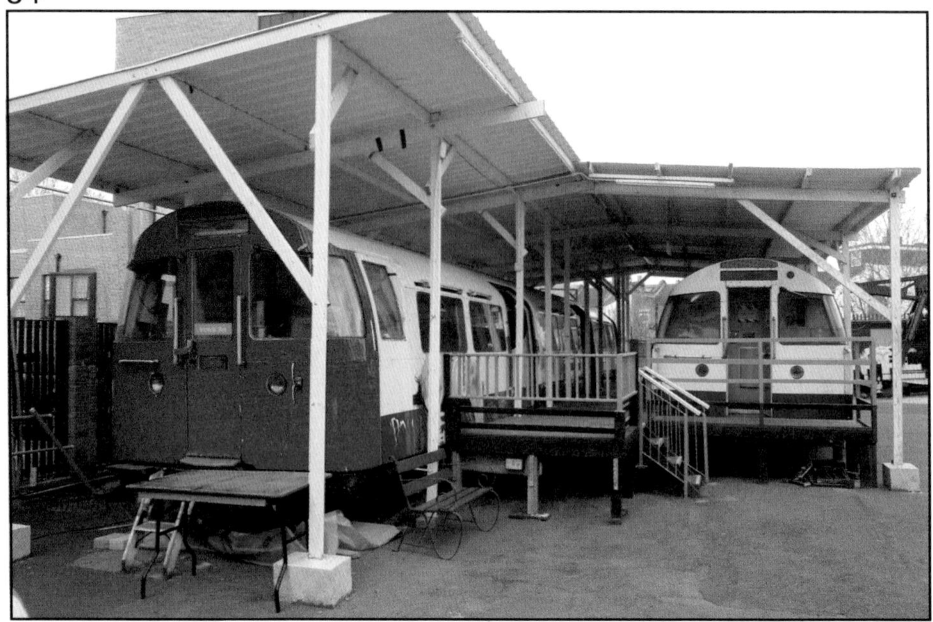

▲ Former Victoria Line 1967 Tube Stock DM 3186 is preserved at the Walthamstow Pumphouse Museum in North-East London where it is seen on 14 April 2019. The vehicle on the right is not complete – it is two parts of former 1967 Tube Stock cars 3049 and 3016. **Robert Pritchard**

▼ 1972 Stock DM 3530 is seen at the London Transport Museum Depot, Acton on 27 April 2019. **Alan Yearsley**

1960 TUBE STOCK — LONDON TRANSPORT EXECUTIVE

Built: 1960. Prototype stock intended for the Central Line. The type never went into series production.
Traction Motors:4 x DC nose suspended GEC LT113 of 49 kW (60 hp).

DM 15.86 x 2.60 m 29.89 tonnes 40 seats.

3906	DM	Cravens Heritage Trains, Epping	Cravens 1960
3907	DM	Cravens Heritage Trains, Epping	Cravens 1960

3906 and 3907 are based at London Underground, Northfields Depot.

A60 STOCK — LONDON TRANSPORT EXECUTIVE

Built: 1960 for the Metropolitan Line.
Traction Motors: 2 x GEC LT114 of 48 kW (65 hp).

DM 16.17 x 2.95 m 31.60 tonnes 54 seats + 4 tip-ups.

5008–5034	DM	London Transport Museum Depot, Acton	Cravens 1961

1962 TUBE STOCK — LONDON TRANSPORT EXECUTIVE

Built: 1962 for the Central Line.
Traction Motors: 2 x GEC LT112 of 60 kW (80 hp).

DM 15.90 x 2.60 m 26.62 tonnes 42 seats.
M 15.60 x 2.60 m 24.28 tonnes 40 seats.

1506	DM	Cravens Heritage Trains, Epping	BRCW 1962
1507	DM	Cravens Heritage Trains, Epping	BRCW 1962
2506	T	Cravens Heritage Trains, Epping	BRCW 1962
9507	M	Cravens Heritage Trains, Epping	BRCW 1962

Based at London Underground, Hainault Depot.

1967 TUBE STOCK — LONDON TRANSPORT EXECUTIVE

Built: 1967–69 for the Victoria Line.
Traction Motors: 2 x Crompton Parkinson LT115 of 53 kW (70 hp).

DM 16.09 x 2.64 m 28.50 tonnes 40 seats.

3052	DM	London Transport Museum Depot, Acton	MC 1968
3122–3186	DM	Walthamstow Pumphouse Museum	MC 1967

1972 TUBE STOCK — LONDON TRANSPORT EXECUTIVE

Built: 1972 for the Northern Line.
Traction Motors: 2 x Brush LT115 of 53 kW (70 hp).

DM 16.09 x 2.64 m 28.20 tonnes 40 seats.

3530	DM	London Transport Museum Depot, Acton	MC 1972
4079	T	*Location unknown*	MC 1972
4179	T	*Location unknown*	MC 1972

4079 and 4179 are owned by Village Underground, Shoreditch and are currently stored at an unknown location in Essex.

C77 STOCK — LONDON TRANSPORT EXECUTIVE

Built: 1977 for the District Line.
Traction Motors: 2 x Brush LT117 of 68 kW (90 hp).

DM 16.03 x 2.92 m 31.70 tonnes 32 seats.

5721	DM	London Transport Museum Depot, Acton	MC 1977

D78 STOCK — LONDON TRANSPORT EXECUTIVE

Built: 1979–83 for the District Line.
Traction Motors: 4 x DC nose-suspended Brush LT118 of 49 kW (65 hp) per motor car.

DM 18.37 x 2.85 m 27.4 tonnes 42 seats + 2 tip-ups.

7012	DM	London Transport Museum Depot, Acton	MC 19??
7027	DM	Coopers Lane Primary School, Grove Park	MC 19??

1983 TUBE STOCK — LONDON TRANSPORT EXECUTIVE

Built: 1983–88 for the Jubilee Line.
Traction Motors: 2 x Brush LT122 of 43 kW (55 hp).

DM 17.23 x 2.63 m 26.34 tonnes 48 seats.

3662	DM	Village Underground, Shoreditch	MC 1988
3721	DM	Tyne & Wear Fire Service Training Centre, Washington	MC 1984
3733	DM	Village Underground, Shoreditch	MC 1988
3734	DM	London Transport Museum Depot, Acton	MC 1988
4633	T	Village Underground, Shoreditch	MC 1983
4662	T	Village Underground, Shoreditch	MC 1983

1986 PROTOTYPE TUBE STOCK — LONDON TRANSPORT EXECUTIVE

Built: 1986–87 for Central Line.
Traction Motors:

DM 16.85 x 2.75 m 44.5 tonnes 30 seats.

16	DM	London Transport Museum Depot, Acton	MC 1986

2.6. SOUTHERN RAILWAY EMU STOCK

The following vehicle was not originally built for London Underground and was used on the Waterloo & City Line when it was operated by Southern Railway (and later BR).

CLASS 487 — SOUTHERN RAILWAY W&C LINE

This vehicle is designated DMBSO – Driving Motor Brake Standard Open.

Built: 1940 for the Southern Railway Waterloo & City Line.
Traction Motors: 2 x EE 500 of 140 kW (185 hp).

DMBSO 14.33 x 2.64 m 29 tons 40 seats.

S 61 S–61	DMBSO	London Transport Museum Depot, Acton	EE 1940

2.7. HAULED COACHING STOCK

The following vehicles were built for London Transport's predecessor companies and are classed as preserved. Vehicles marked * are cleared to run on London Underground metals and see occasional use on specials.

CITY & SOUTH LONDON RAILWAY "PADDED CELL" CAR

These vehicles were nicknamed "Padded Cells" because of their high-backed seats with small windows above them.
Built: 1890 for the City & South London Railway.
Wheel arrangement: 2 x 4-wheel bogie.
Layout: Saloon with longitudinal seating.
Condition: Fully restored static exhibit.

9.75 m x ?? m   7 tons   32 seats.

30	London Transport Museum	Ashbury 1890

CITY & SOUTH LONDON RAILWAY BRUSH STOCK

Two examples of these vehicles can be classed as preserved. However, only their bodies survive.
Built: 1902 (135), 1907 (163) for the City & South London Railway.
Wheel arrangement: Not applicable (body only).
Layout: Saloon with longitudinal seating.
Condition: Body only without interior.

135	Hope Farm, Sellindge, Kent	Brush 1902
163	London Transport Museum Depot, Acton	Brush 1907

METROPOLITAN DISTRICT RAILWAY 4-WHEEL FIRST

Built as a First Class coach, but now used as a Third. This vehicle is mounted on the chassis of Southern Railway design PMV 1225, with only its body surviving. After withdrawal its body was moved to Dymchurch where it was used as a store shed. Restored by Resco Railways, Woolwich 1976–80.
Built: 1884 for the Metropolitan District Railway.
Wheel arrangement: 4-wheel.
Type: Third (originally First).
Layout: Non-corridor compartments.
Number of compartments: 4.
Condition: Operational.

100*	Kent & East Sussex Railway	Ashbury 1884

METROPOLITAN RAILWAY "JUBILEE" STEAM STOCK

This vehicle is mounted on the chassis of Southern Railway design PMV 1647, with only its body surviving. 353 had been sold to the Weston, Clevedon & Portishead Railway in 1906. It was withdrawn in 1940 and its body moved to Shrivenham for use as a tailor's shop. Restored by Ffestniog Railway, Boston Lodge 2011/12 with support from the LT Museum in time for the 150th anniversary of the Underground in January 2013.
Built: 1892 for the Metropolitan Railway.
Wheel arrangement: 4-wheel.
Type: First.
Layout: Non-corridor compartments.
Number of compartments: 4.
Condition: Operational.
8.5 m x 2.5 m

MR	*WCPR*		
353*	12	Kent & East Sussex Railway	Cravens 1892

On loan from the London Transport Museum.

METROPOLITAN RAILWAY MILK VAN

Built: 1896 for the Metropolitan Railway. Converted for use as a Breakdown Train Tool Van by the LPTB. Restored for the 1963 London Underground centenary celebrations.
Wheel arrangement: 4-wheel.
Condition: Fully restored static exhibit (in operational condition, occasionally used on specials).

Type	*Length x width*
MV	8.2 m x 2.6 m

MR	*LTPB*		
3*	BDV700	London Transport Museum Depot, Acton	BRCW 1896

METROPOLITAN RAILWAY "CHESHAM" STOCK

These vehicles were built as hauled stock designated "BOGIED" and converted for electric operation in 1906 (368/387/412) and 1921 (394/400). 368/412 were designated as "M" stock. 387 was converted to a DBT for use in a push-pull set, then in 1908 was further converted to a DMBT and designated as "N" Stock. 394/400 were converted to DTT and designated as "W" stock. All were further converted for steam operation on the Chesham branch in 1940, 512 being converted to a DBT.

Type	*Length x width*	*Weight*	*Seats (compartments)*
C	12.04 m x 2.5 m	20 tons	1st 24(3), 3rd 30(3)
DBT	12.04 m x 2.5 m	20 tons	40(4) (originally 50(5))
DTT (converted from T)	12.04 m x 2.5 m	20 tons	60(6) (originally 70(7))
DTT§ (converted from S)	12.04 m x 2.5 m	19 tons 10 cwt	60(6) (originally 70(7))

Wheel arrangement: 2 x 4-wheel bogie.
Layout: Non-corridor compartments.
Condition: Operational (except 400: fully restored static exhibit).

MR	*LTPB*	*LTE*			
368*	9702	515	C	Bluebell Railway	Ashbury 1898
387*	2761	512	DBT	Bluebell Railway	Cravens 1900
394*	6702	518	DTT	Bluebell Railway	Ashbury 1900
400†	6703	519	DTT§	London Transport Museum	Neasden 1900
412*	9705	516	C	Bluebell Railway	Cravens 1900

† Car 400 has compartments furnished to represent different time periods: one in as-built condition, a "Ladies Only" compartment in 1930s style, and the rest in 1950s style.

§ Car 400 was originally a Second Class car. Second Class was abolished on the Metropolitan Railway c1905 and all such accommodation was then redesignated as Third Class.

METROPOLITAN RAILWAY "DREADNOUGHT" STOCK

Built: 1910 (427), 1919 (465), 1923 (509) for the Metropolitan Railway.
Wheel arrangement: 2 x 4-wheel bogie.
Layout: Non-corridor compartments.
Condition: Operational (normally kept on display as static exhibits).
These vehicles are owned by the Vintage Carriages Trust.

Type	*Length x width*	*Weight*	*Seats (compartments)*
BT	16.56 x 2.82 m	30 tons	84(7)
F	16.56 x 2.82 m	30 tons	84(7)
T	16.56 x 2.82 m	30 tons	108(9)

427	BT	Keighley & Worth Valley Railway	MC Birmingham 1910
465	T	Keighley & Worth Valley Railway	MC Birmingham 1919
509	F	Keighley & Worth Valley Railway	MC Birmingham 1923

▲ City & South London Railway "Padded Cell" coach No. 30 (part of set No. 10) at the London Transport Museum, Covent Garden on 6 April 2019. Opening in 1890 the City & South London Railway was the world's first Underground electric railway. The stock had only very narrow windows as it was thought there would be little for passengers to see in the tunnels. **Robert Pritchard**

▼ Metropolitan Railway "Chesham" stock DTT coach 400 at the London Transport Museum, Covent Garden on 6 April 2019. **Robert Pritchard**

3. UK LIGHT RAIL & METRO SYSTEMS

After a period of almost total abandonment in the decades following World War II, tramways have enjoyed a renaissance in the UK since the 1990s. This section contains details of the rolling stock fleets of all operational UK tram, light rail and metro systems. In addition to the seven tram networks, there are also three light metros in the UK: the Docklands Light Railway, the Glasgow Subway and the Tyne & Wear Metro. These are classed as metros rather than tramways because they run entirely on segregated rights of way rather than on-street. Heritage tramways such as those at Birkenhead and Seaton, and museum sites such as Beamish and Crich, are not included in this book. The heritage tram fleet at Blackpool has been included, however, as the majority of the vehicles are cleared to operate on a public tramway.

The UK had the honour of having the world's first passenger carrying railway, the Swansea & Mumbles Railway, opened to passengers in 1807. This was also the world's first passenger tramway because, although referred to as a railway it used tramway type vehicles throughout its life, starting with horse-drawn trams. It would be several more decades before other towns and cities followed suit, with the next horse tramway opening in 1860 in Birkenhead. Many more horse tramways followed in the latter part of the 19th century, and by the 1870s and 1880s some of these were beginning to give way to steam traction. The late 19th and early 20th century also saw some towns and cities such as Edinburgh and Matlock experiment with cable propelled tramways following the example set by San Francisco, USA, with its world-famous cable tramway which opened in 1873. The Glasgow Subway also used the cable system from its opening in 1896 until 1935. Today, there are no surviving street running cable tramways in the UK apart from the Great Orme Tramway in Llandudno, North Wales. However, strictly speaking this is a funicular rather than a cable tramway, as the cars are stopped and started by stopping and starting the cable, to which they are permanently fixed, unlike a cable tramway where the cable runs continuously and the cars can be attached to and detached from the cable. The Great Orme Tramway is not included in this book, as it is classed as a heritage/ tourist tramway.

Blackpool was the UK's first electric tramway, opening in 1885 following earlier examples of electric traction in the early 1880s such as Sestroretsk in Russia, Lichterfelde, near Berlin, and the Mödling and Hinterbrühl Tram in Austria. Blackpool initially used a conduit electrification system whereby trams collected electric current from a conduit beneath the tracks via a groove between the two running rails. This system was also used on tram lines in central London, but soon proved to be unsuitable for Blackpool because the conduit was frequently damaged by sand and sea water, and it was replaced by overhead wires in 1899.

By the early 1900s, a small number of horse tramways had closed, but most had been converted to electric traction. This period also saw the opening of several completely new electric tramways, including many in small towns such as Merthyr Tydfil and Weston-super-Mare.

The heyday of the electric tram in the UK can perhaps best be described as having been between the 1900s and the 1920s. Sheerness & District Tramways was one of the shortest-lived tram systems in the country, running only from 1903 until 1917. Tramway closures began in earnest in the late 1920s, with several more small and medium-sized town networks being abandoned in the 1930s. By this time tramways were increasingly being replaced by buses and trolleybuses. Significant pre-war closures included Wolverhampton in 1928 and Nottingham in 1936.

After World War II, Manchester was the first major city to abandon its tram network in 1949, with almost all remaining systems succumbing during the 1950s such as Edinburgh in 1956, Liverpool in 1957, Aberdeen in 1958 (despite having introduced a new fleet of streamlined trams only 9 years earlier) and Leeds in 1959. By this time rising levels of car ownership and improvements in the design of motor buses meant that tramways were falling out of favour in the UK and in many other countries such as Canada, France, the USA and Australia. The last two tramway closures in the UK took place in Sheffield in 1960 and Glasgow in 1962, after which Blackpool became the sole surviving tramway in the country.

Increasing road congestion led the Government and local authorities to recognise the need for more radical public transport-based solutions. The UK's first modern light rail system was the Tyne & Wear Metro, opened in 1980, followed by the Docklands Light Railway in 1987. Manchester led the way in bringing trams back to the UK with the opening of its Metrolink system in 1992, and has since been joined by Sheffield, Birmingham, Croydon, Nottingham and Edinburgh. A number of other tramway schemes, including the Bristol and Leeds Supertram and Liverpool's Merseytram, have been proposed but abandoned.

Since their opening, the Manchester and Nottingham tramways have expanded considerably (as have the Docklands Light Railway and to a lesser extent the Tyne & Wear Metro) and are now much larger than their original networks. The Sheffield system also expanded in 2018 with the addition of tram-train service to Rotherham Parkgate and the West Midlands Metro is also currently expanding significantly. Edinburgh will be extended to Newhaven soon but the tramway in Croydon has so far remained unchanged since opening.

Despite the revival in the fortune of tramways in the UK in recent decades, only six new tram networks have so far opened. This is modest compared to some other European countries such as France where 22 new tramways have opened since the 1980s (with only three tram systems having operated continuously since opening in the late 19th century: Lille, Marseille and Saint-Étienne), starting with Nantes in 1985. There are also two French cities with rubber tyred vehicles guided by a single rail: Clermont-Ferrand and Nancy (Caen having also operated such a system until the end of 2017, but this has now been converted to a conventional tramway which is expected to open in the autumn of 2019).

Trams have traditionally operated with a conductor who would issue and check tickets. Currently Birmingham, Blackpool and Sheffield trams have conductors, and on all other tramways and light rail networks tickets must be purchased before boarding and may be checked by ticket inspectors during the journey or when boarding or alighting. Information on ticketing and fares is included in the entry for each tramway.

To date, no two tram or light rail networks in the UK have used an exactly identical design of vehicle when first opening, with each system opting for a completely new design; however, West Midlands Metro's second fleet of trams, the CAF Urbos 3, is similar to those in Edinburgh. Also, many of the vehicle fleets are based on an existing design from abroad: London Tramlink's Bombardier Flexity Swift CR4000s are modelled on the K4000s in Cologne, Germany, and the Stadler Variobahns are similar to those used by Bybanen in Bergen, Norway. Similarly, Nottingham Express Transit's Bombardier Incentro AT6/5 was based on those built for Nantes, France.

▲ **Tram-train 399 201 arrives at the low level platforms at Rotherham Central with a Supertram service from Parkgate to Cathedral (Sheffield) on 7 February 2019. Robert Pritchard**

3.1. BLACKPOOL & FLEETWOOD TRAMWAY

Network: 11 miles. **Lines**: 1. **Depots**: 2 (Starr Gate and Rigby Road). **First opened**: 1885.
www.blackpooltransport.com **System**: 600 V DC overhead.
Standard livery: Flexity 2s and **F**: White, black & purple.

Until the opening of Manchester Metrolink, the Blackpool tramway was the only urban/inter-urban tramway system left in Britain. As mentioned in our Introduction, the seaside resort of Blackpool was the location for the UK's first electric tramway, opening on 29 September 1885 and operated by Blackpool Corporation from 1892. In 1898, the Blackpool & Fleetwood Tramroad Company opened a rival line between Gynn Square and Fleetwood Ferry. The two tramways were not connected until the Fleetwood line was taken over by Blackpool Corporation in 1920, enabling through running between South Shore and Fleetwood. By the 1920s there was an extensive network of street tramways to Layton, Squires Gate and Marton. These had all closed by 1962 so trams were concentrated on the Starr Gate–Fleetwood section. While all other UK tramways had closed by 1962, Blackpool stayed open, partly because of its unique value as a tourist attraction in its own right.

Today, Blackpool has a population of around 140 000. The tramway consists of a single route, 11 miles long, from the town of Fleetwood on the Fylde coast to the north of Blackpool, through Cleveleys and Bispham to Blackpool itself, where trams run along the promenade and beneath the famous Blackpool Tower, which is visible for miles around. After Pleasure Beach the line continues to the southern terminus at Starr Gate.

The Blackpool tramway is municipally operated by Blackpool Transport Services Ltd, which is owned by Blackpool Council and also runs buses in Blackpool. The infrastructure is owned by Blackpool Corporation.

After years of under-investment and a fleet of trams mainly dating mainly from the 1930s, a decision was made in 2008 to upgrade the tramway to a modern light rail system, with new low-floor trams and rebuilt tracks and stops, at a cost of around £100 million. The system closed for upgrading in November 2011, and reopened in April 2012 with a new fleet of Bombardier Flexity 2 trams: these are now used on the normal scheduled services. Part of the upgrade also included a new state-of-the-art depot at Starr Gate – the original depot at Rigby Road depot has been retained to maintain the "B" fleet (see below) and heritage trams.

The Heritage aspect of the operation has not been forgotten, and still brings in significant revenue for the tramway, especially the "illuminations tours" which operate every autumn in connection with the famous Blackpool Illuminations. The heritage side of the operation is operated by Blackpool Heritage Tram Tours, part of Blackpool Transport Services.

Today the fleet is split into three distinct fleets: The "A" fleet comprises 18 new Flexity cars used for the normal service. These Flexity 2 trams were built by Bombardier in Germany and are 5-section trams riding on three bogies, the two outer bogies being motored. The first 16 were delivered in 2011–12 and were followed by two further cars (017 and 018) which were delivered in December 2017 to cover the planned service on the Blackpool North station extension. The Flexity 2 trams have a total capacity (standing and seating) of 200 passengers.

To supplement the Flexity service it was planned to retain and refurbish nine of the 1930s-built double-deck "Balloon" trams to operate a supplementary service between Pleasure Beach and Cleveleys at busy times. Nine Balloon cars were modified for this purpose and have a partial exemption from the Rail Vehicle Accessibility Regulations (RVAR) for use on normal services, forming the "B" fleet. This is because they have been made more accessible than in the past, and also on the basis that a fully compliant tram will always be available within 5 minutes of a "B" fleet car. The Balloons have been fitted with wide centre sliding plug doors to allow them to call at the rebuilt platforms. Despite the money spent on upgrading the fleet, the refurbished Balloon cars have seen little use, with the Flexity trams able to generally adequately cover the normal service and any extras required to operate.

Finally, the "C" fleet is the Heritage Fleet of Vintage cars, which have full exemption from the RVAR regulations to operate separate heritage, open-top or illuminated services. Initially there were planned to be 16 vehicles to form the core "C" fleet: Box car 40, Bolton 66, Standard 147, Open Boat cars, 600 and 602 and 604, Brush car 631, Centenary car 648, Corporation car 660, Progress twin-cars 672+682 and 675+685, Open-top Balloon 706, Balloon 717, Illuminated Western Train 733+734 (which is fitted with wheelchair access), Illuminated Warship 736, Illuminated Trawler 737. These have since been augmented by other former Blackpool trams

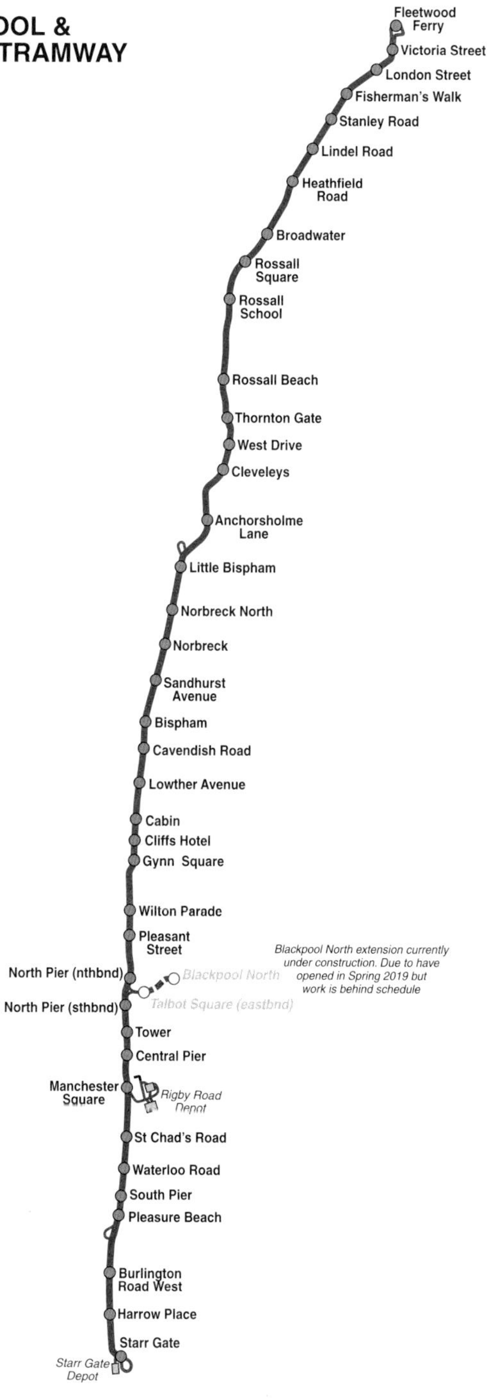

BLACKPOOL &
FLEETWOOD TRAMWAY
Fleetwood Ferry
Victoria Street
London Street
Fisherman's Walk
Stanley Road
Lindel Road
Heathfield Road
Broadwater
Rossall Square
Rossall School
Rossall Beach
Thornton Gate
West Drive
Cleveleys
Anchorsholme Lane
Little Bispham
Norbreck North
Norbreck
Sandhurst Avenue
Bispham
Cavendish Road
Lowther Avenue
Cabin
Cliffs Hotel
Gynn Square
Wilton Parade
Pleasant Street
North Pier (nthbnd)
Blackpool North
Blackpool North extension currently under construction. Due to have opened in Spring 2019 but work is behind schedule
North Pier (sthbnd)
Talbot Square (eastbnd)
Tower
Central Pier
Manchester Square
Rigby Road Depot
St Chad's Road
Waterloo Road
South Pier
Pleasure Beach
Burlington Road West
Harrow Place
Starr Gate
Starr Gate Depot

that have joined the fleet, and visitors from preserved tramways such as the National Tramway Museum at Crich or the North of England Open Air Museum at Beamish. Box car 40 has now been returned to Crich.

In terms of services, starting with the normal service, this operates either every 10 or 15 minutes during the day, depending on the time of year, with the journey from Fleetwood to Starr Gate timetabled to generally take 55 minutes. In the winter the service operates every 15 minutes between 08.00 and 19.00 (every 30/20 minutes in the early morning and every 30 minutes in the evening), requiring eight trams. In the peak season (April–October) the service runs every 10 minutes, being allowed 58 minutes end-to-end and requiring 13 trams. Additional short workings (such as Pleasure Beach–Little Bispham) can also operate at time of high demand.

The Heritage operation was scaled back for the 2014 and 2015 seasons, but expanded again from 2016. From early January until early November at least two cars from the heritage fleet are in operation between Pleasure Beach and North Pier or Cabin, operating to a frequent timetable (usually every 20–30 minutes between 10.00 and 17.00). The Heritage service also operates on weekdays during school holidays and every day in July and August and between mid October and early November. Santa specials normally operate on weekends in December. Once a month an enhanced "Gold" heritage services operates (often on Bank Holiday weekends) with six or more heritage trams in operation, including trips the full length of the line to Fleetwood. Dates of these weekends are published on the Blackpool transport website or on the dedicated Heritage operation site – www.blackpoolheritage.com. The Heritage services call at slightly different stops from normal services – there are six that can be served by the Heritage services (Pleasure Beach, Tower/North Pier, Cabin, Bispham, Cleveleys and Fleetwood).

During the illuminations season (early September until early November) tram tours using Heritage or Illuminated vehicles operate from Pleasure Beach to Bispham from dusk every evening. Boarding of these special trips is only allowed at Pleasure Beach, but passengers may alight on the return at Bispham, Cabin or Tower/North Pier. A Heritage day ticket is valid to travel on the Illuminations tours.

It should be noted that unlike the other tramways listed in this book Blackpool operates from a lower voltage – of 600 V DC. This was increased from 550 when the tramway was upgraded in 2011–12. Trams are restricted to 30 km/h (around 19 mph) along the promenade as far as Cabin, and also through the streets of Fleetwood.

Tickets can be purchased from conductors on the trams in the time-honoured way. Single fares start from £1.80. 24 hour "Saver" tickets valid on buses and trams in Blackpool cost £5.50. 3 and 7 day tickets are also available. A Heritage ticket giving day travel on the heritage services (if operating) as well as the normal tram and bus services costs £11 and can be bought on board heritage tram services or on normal service trams. A round-trip ticket for use on the heritage trams is also available for £3.50 (Pleasure Beach–Cabin) or £5.50 (Pleasure Beach–Fleetwood).

According to the Department for Transport annual light rail usage figures Blackpool trams carried 5.2 million passengers in 2018–19. This illustrates the success of the upgraded tramway and the enhanced capacity the Flexity trams provide, being the highest figure for more than 20 years. These figures exclude passengers carried on the heritage services, which are counted separately.

Work started in 2018 of an extension of the tramway from North Pier up Talbot Road to Blackpool North station. It was originally planned that the short extension would open in spring 2019, but there have been a number of delays, not least with the planned demolition of a Wilkinson store that lies in the path of the planned route of the tramway near Blackpool North station. At the time of going to press the new extension was expected to open by the end of 2020. Local campaign group Trams to Lytham is also backing an extension of the tram network south to St Anne's and Lytham but this is subject to further feasibility studies and development work.

▲ The 18 Bombardier Flexity 2 cars work all normal services on the Blackpool Tramway. 002 "Alderman E.E. Wynne" arrives at Starr Gate on 7 November 2018. **Robert Pritchard**

▼ 018, one of the two additional trams which entered service in 2018, is seen leaving North Pier with a service for Starr Gate on 16 September 2018. **Tony Christie**

FLEXITY 2 "A FLEET" 5-SECTION TRAMS

The Flexity 2 trams operate normal scheduled services. Based at Starr Gate depot.
Built: 2011–12 or 2017 (017/018) by Bombardier Transportation, Bautzen, Germany.
Wheel Arrangement: Bo-2-Bo.
Traction Motors: 4 x Bombardier 3-phase asynchronous of 120 kW (161 hp).
Seats: 70 (+4). **Weight**: 40.9 tonnes.
Dimensions: 32.20 x 2.65 m. **Doors**: Sliding plug.
Braking: Regenerative, disc and magnetic track.
Couplers: **Maximum Speed**: 43 mph.
Advertising Livery: 003 and 016 – Prettylittlething.com (pink).

001
002
003 **AL**
004
005
006
007
008
009
010
011
012
013
014
015
016 **AL**
017
018

Names:

002 Alderman E.E. Wynne
007 Alan Whitbread

"BALLOON" DOUBLE DECKERS "B FLEET" A1-1A

The nine cars listed have partial exemption from the Rail Vehicle Accessibility Regulations (RVAR) and have been fitted with wider doors. They see limited use, often during the autumn illuminations season or in high summer. They are also sometimes used on Heritage tours. Based at Rigby Road depot.

Built: 1934–35 by English Electric, Preston.
Traction Motors: 2 x EE305 of 40 kW (53 hp). **Seats**: 94 (*† 92).
Advertising Liveries: 709: Blackpool Sealife Centre (blue).
713: Houndshill Shopping Centre (purple & white).
720: Walls ice cream (red).
724: Lyndene Hotel (blue).

* Rebuilt with a flat front end design and air-conditioned cabs.

700 **F**
707 * **Green/cream**
709 * **AL** (S)
711 † **F**
713 **AL**
718 * **Green/cream**
719 **F**
720 **AL** (S)
724 * **AL** (S)

Names:

711 Ray Roberts
719 Donna's Dream House

HERITAGE FLEET of VINTAGE CARS "C FLEET"

The following trams have exemption from the Rail Vehicle Access Regulations (RVAR) for Heritage use. They normally see use on the weekend heritage trips, during the autumn Illuminations season or for private excursions. Some vehicles are undergoing restoration. Based at Rigby Road depot.

Bolton 66[1]	Bogie double-decker	Built: 1901
Blackpool 143	Standard double-decker	Built: 1924
Blackpool 147 "MICHAEL AIREY"	Standard double-decker	Built: 1924
Blackpool 225–600 "THE DUCHESS OF CORNWALL"	Open boat car	Built: 1934
Blackpool 227–602 "CHARLIE CAIROLI"	Open boat car	Built: 1934
Blackpool 230–604 "GEORGE FORMBY OBE" (S)	Open boat car	Built: 1934
Blackpool 284–621	Brush Railcoach	Built: 1937
Blackpool 294–631	Brush Railcoach	Built: 1937
Blackpool 304–641	Coronation Class single-decker	Built: 1952
Blackpool 642	Centenary car	Built: 1986
Blackpool 648	Centenary car	Built: 1985
Blackpool 272+T2–672+682 (S)	Progress Twin Car	Rebuilt: 1960
Blackpool 275+T5–675+685	Progress Twin Car	Rebuilt: 1958/60
Blackpool 280–680[2]	English Electric Railcoach	Rebuilt: 1960

▲ Interior of Flexity 2 car 017. **Robert Pritchard**

▼ Red & Cream-liveried Open boat car 227 is seen reversing at Cabin with one of the Heritage services on 15 June 2019. **Alan Yearsley**

Blackpool 238–701	Balloon double-decker	Built: 1934
Blackpool 240–703	Balloon double-decker	Built: 1934
Blackpool 243–706 "PRINCESS ALICE" (S)	Balloon open-top double-decker	Built: 1934
Blackpool 252–715	Balloon double-decker	Built: 1935
Blackpool 254–717 "WALTER LUFF"	Balloon double-decker	Built: 1935
Blackpool 260–723	Balloon double-decker	Built: 1935

Illuminated cars

Blackpool 733	Western Train loco & tender	Rebuilt: 1962
Blackpool 734	Western Train carriage	Rebuilt: 1962
Blackpool 736	"Frigate" HMS Blackpool	Rebuilt: 1965
Blackpool 737	"Fisherman's Friend" Illuminated Trawler	Rebuilt: 2001

[1] Owned by the Bolton 66 Tramcar Trust.
[2] On loan from the Manchester Transport Museum Society, Heaton Park Tramway.

STORED VEHICLES

The following Heritage vehicles are currently stored at Rigby Road depot.

Blackpool 8	BCT One Man Car	Rebuilt: 1974
Lytham 43	English Electric double-deck open balcony car	Built: 1924
Blackpool 279–679	English Electric Railcoach (ex-Twin Car motor)	Rebuilt: 1960
Blackpool 287–259–624	Brush Railcoach	Built: 1937
Blackpool 288–625[3]	Brush Railcoach	Built: 1937
Blackpool 290–627[4]	Brush Railcoach	Built: 1937
Blackpool 295–632	Brush Railcoach	Built: 1937
Blackpool 297–634	Brush Railcoach	Built: 1937
Blackpool 645	Centenary car	Built: 1987
Blackpool 324–660	Coronation Class single-decker	Built: 1953
Blackpool 327–663[5]	Coronation Class single-decker	Built: 1953
Blackpool 271+T1–671+681	Progress Twin Car	Rebuilt: 1960
Blackpool 276+T6–676+686	Progress Twin Car	Rebuilt: 1958/60
Blackpool 241–704	Balloon double-decker	Built: 1934
Blackpool 245–708[6]	Balloon double-decker	Built: 1934
Blackpool 263–726[3]	Balloon double-decker	Built: 1935
Blackpool 168–732	Rocket illuminated car	Built: 1961
Blackpool 222–735	Hovertram illuminated car	Rebuilt: 1963
Blackpool 761	Jubilee Class double-decker	Rebuilt: 1979
Halle 902[7]	Tatra T4D	Built: 1969
Glasgow 16–1016	Open-top double decker	Built: 1904
CityClass 611	Trampower Prototype articulated car	Built: 1997

[3] Privately owned and at Rigby Road for storage/private restoration only.
[4] Owned by the Fleetwood Heritage Leisure Trust and at Rigby Road for storage only.
[5] Currently at Ian Riley Ltd, Heywood for restoration.
[6] Owned by the Manchester Transport Museum Society.
[7] Stored on behalf of Crich Tramway Village.

ENGINEERING CAR

Built: 1992 as a purpose-built overhead line engineering car for Blackpool Transport, the first completely new works car on the system. Fitted with an inspection platform.

754	Rebuilt: 1992

▲ Recently returned to service, superbly restored Brush Railcoach 621 is seen at North Pier with one of the Heritage services to Pleasure Beach on 16 September 2018. **Tony Christie**

▼ Balloon double-decker 717 is seen at Pleasure Beach with an Illuminations special on 3 November 2018. **Alisdair Anderson**

3.2. DOCKLANDS LIGHT RAILWAY

Network: 23½ miles. **Lines**: 6. **Depots**: 2 (Beckton and Poplar). **First opened**: 1987.
www.tfl.gov.uk/dlr **System**: 750 V DC third rail (bottom contact).

The unique Docklands Light Railway in east London is a driverless Metro where trains are normally driven automatically using the Thales Seltrac moving block signalling system. They can also be driven manually by the "Passenger Service Agent", the member of staff present on every train.

First conceived in 1981 as a link between the city of London and new offices in the regenerated Isle of Dogs and Canary Wharf areas, the first part of the DLR system opened in summer 1987 with two routes. One linked Tower Gateway (close to Fenchurch Street) with Island Gardens via Canary Wharf, while the other ran from Stratford to Island Gardens. This initial network of 8 miles with 15 stations used just 11 of the original P86 trains, running singly. These were joined by ten P89 trains in 1989. The original routes were largely on old railway routes, with Tower Gateway–Island Gardens mostly on viaducts, while the later extensions have mostly new formations. Platforms were soon extended and 2-car operation started in 1991 when the first of the B90 cars were delivered.

Originally owned by London Transport, DLR is now part of the London Rail division of Transport for London (TfL), operated by a consortium of Keolis and Amey on behalf of TfL. KeolisAmey Docklands is contracted to run DLR from December 2014 to April 2021 with the option for this to be extended to 2023. KeolisAmey took over from Serco, which had operated DLR since 1997.

In the 30 years since opening DLR has gone from strength to strength and is now an integral part of the London public transport network. With 149 trains compared to 11 when it opened, it runs between two termini in central London, at Bank and Tower Gateway, to Lewisham, Stratford/Stratford International, Beckton and Woolwich Arsenal.

The dates of the numerous extensions are given below. Some of these lines used former railway alignments. Tower Gateway proved inadequate as a London terminus and a new underground station at Bank (accessed via twin tunnels and a 1 in 17 gradient), more convenient for onward Underground connections and the city of London, opened in 1991. The 1994 extension to Beckton made extensive use of viaducts and also serves the ExCel exhibition centre. It also gives access to a new second depot, relieving pressure on the original cramped depot at Poplar.

A further extension south to Lewisham 5 years later saw DLR cross to the southern side of the River Thames for the first time. The original elevated Island Gardens station was replaced with a new sub-surface station when the Lewisham extension opened. In the 2000s Woolwich Arsenal, via London City Airport and a second crossing of the river was reached via an extension from Canning Town. The Stratford International route partly replaced former North London Line services between Canning Town and Stratford. Opening dates for the DLR lines are as follows:

31 August 1987: Tower Gateway–Island Gardens/Poplar–Stratford;
29 July 1991: Shadwell–Bank;
28 March 1994: Poplar–Beckton;
20 November 1999: Island Gardens–Lewisham;
2 December 2005: Canning Town–King George V;
10 January 2009: King George V–Woolwich Arsenal;
31 August 2011: Canning Town–Stratford–Stratford International.

Today DLR has 45 stations. Despite the different routes, route numbers or colours are not used. Services are very frequent across the network, combining to give a train every 2 minutes at peak times on the very busy Limehouse–Shadwell section (every 2½ minutes off-peak). Details of the different routes and their frequency on weekdays is given below:

Bank–Canary Wharf–Lewisham: Every 5 minutes/4 minutes (peak). Journey time 25 minutes;
Bank–Poplar–Woolwich Arsenal: Every 10 minutes/8 minutes (peak). Journey time 26 minutes;
Tower Gateway–Poplar–Beckton: Every 10 minutes/8 minutes (peak). Journey time 26 minutes;
Stratford Int–Canning Town–Woolwich Arsenal: Every 10 min/8 min (peak). Journey time 23 min;
Stratford–Poplar–Canary Wharf*: Every 5 minutes/4 minutes (peak). Journey time 13 minutes;
Canning Town–Beckton: Every 10 minutes (off-peak). Journey time 13 minutes.
* Extended to Lewisham in the morning peak.

On weekdays all services the first three services listed normally use 3-car sets, whilst Stratford International–Woolwich Arsenal, Stratford–Canary Wharf/Lewisham and the off-peak Canning Town–Beckton shuttle use 2-cars. On Sundays all services are normally 2-cars apart from the Bank–Lewisham route which is 3-cars.

DOCKLANDS LIGHT RAILWAY

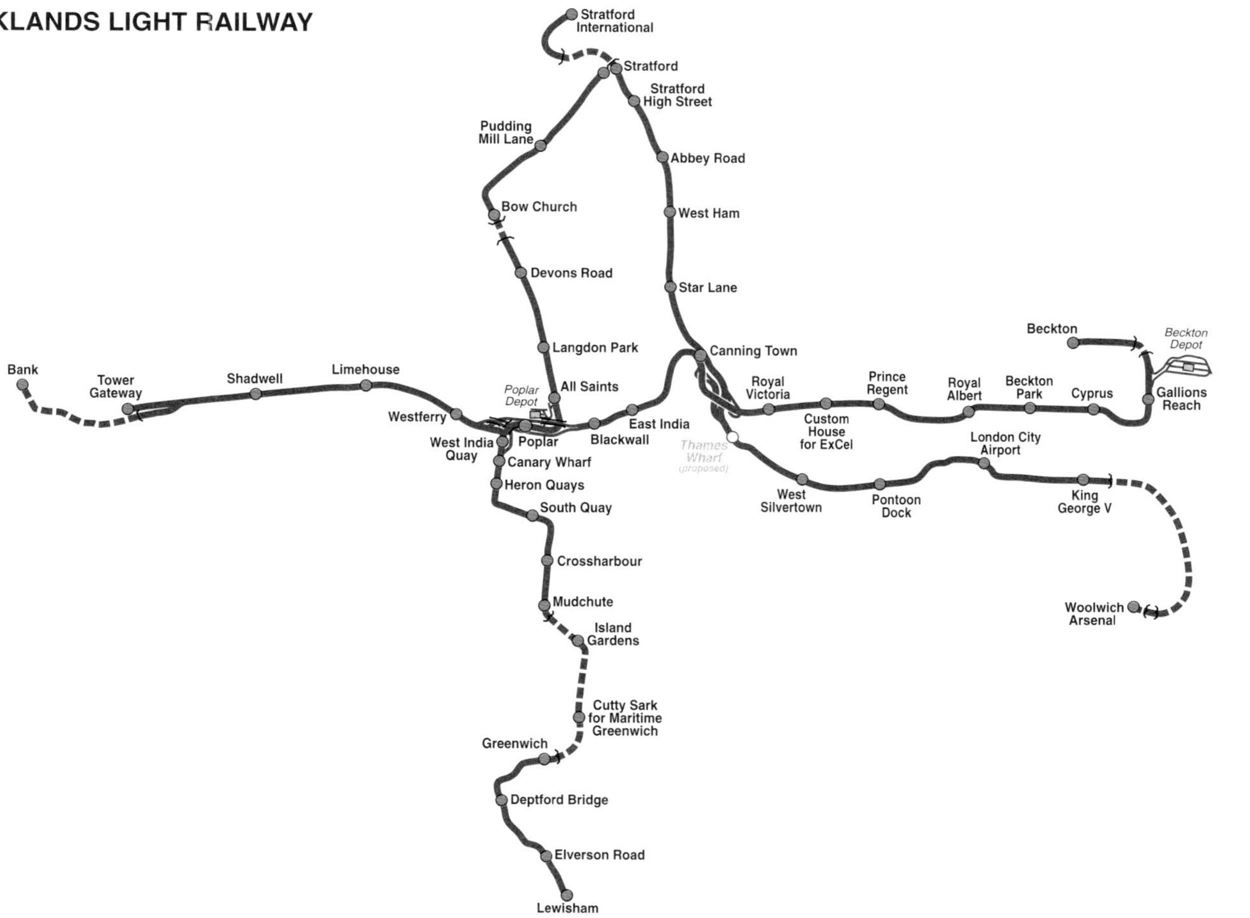

▲ B90/B2K/B92 cars 29, 03 and 76 leave Shadwell with a Lewisham–Bank service on 29 March 2019. **David Palmer**

Trains run from around 05.30 to midnight. An unusual quirk of the timetable is that the Canning Town–Beckton section has a more frequent service in the off-peak than the peak timetable (every 5 minutes off-peak but every 8 minutes peak).

Rather like London Underground stock, the DLR rolling stock class designations (ie "B90") denote the year that class of stock was first designed. The different types of stock are technically very similar, although the newest, B2007 cars look rather different. All trains are articulated 2-car sets, although referred to by DLR as one train. The oldest trains were built by BN Construction of Belgium (a company later taken over by Bombardier). As they were not suitable for underground operation (to Bank), the original P86 and P89 Class vehicles 01–21 were withdrawn from service in 1991 (01–11) and 1995 (12–21) and are now used by the Metro in Essen, Germany. 23 similar B90 cars were delivered in 1991–92, and were followed by 47 B92 cars in 1992–95 – principally for the Beckton extension. The final cars built to the original design were 24 Class B2K cars built 2002–03 (numbered 92–16 these duplicate some of the numbers of the original cars). These were built by Bombardier in Belgium. B90, B92 and B2K cars can all operate with each other, but not with the newer B2007 cars. The vehicles all have a floor height of 1025 mm above rail level and provide step-free access.

To enable 3-car operation 55 more cars were ordered from Bombardier. Called B2007 cars these entered traffic between 2008 and 2010. Technically they are very similar to the earlier cars, being 2-section units with motor bogies at each end. They were built by Bombardier at its Bautzen plant in Germany. 3-train operation started from February 2010, initially between Bank and Lewisham, and now most trains are formed of three units. In 2015 the seating layout in the B2007s was changed to mostly longitudinal (inwards facing) Tube style seating, to create more standing room.

The original livery was blue, red and white. This was replaced from 2004 by a mainly red livery, with a curving blue stripe to represent the River Thames.

The current timetable requires 134 of the 149 units in service, with nominally 140 available for service (with two 3-car "hot spare" trains). There are two depots, the large depot and stabling sidings at Beckton, plus the original depot and workshops at Poplar. At Poplar a diesel-hydraulic shunting locomotive and battery electric works locomotive are based at the depot. The DLR control centre is located at Beckton.

All trains are staffed by Passenger Service Agents, who are responsible for closing the doors and also checking tickets. DLR forms part of the TfL network, with interchange available at many places with London Underground or heavy rail, the main interchanges being at Bank, West Ham, Canning Town, Stratford, Woolwich Arsenal and Lewisham. A single with a contactless card or Oyster smartcard costs between £1.50 and £3.90, with the price being higher at peak times. A Zones 1–4 Travelcard is valid on the whole of the DLR network. The system is generally "open", most stations not having ticket gates, but tickets must be purchased before boarding trains. Oyster and contactless card users must "sign in" at the start and "sign out" at the end of their journey using the yellow card readers.

Given the size of today's network there is not one place where all trains can be seen; however, Poplar is a good location for observing services – and those trains not serving the station can be seen on the viaduct near West India Quay station, so all but trains on the Stratford International–Woolwich and Canning Town–Beckton routes can be seen. For the services running into Tower Gateway and Bank, Shadwell is a good station for observing – and c2c services can also be seen on the adjacent main line. Given the fact that it is largely elevated, DLR has become something of a tourist attraction in its own right, thanks to the fine views that can be had of the changing London landscape – especially from the front windows. DLR also allows easy access to the Emirates Airline cableway, the north bank terminus being close to the Royal Victoria stop.

Patronage continues to increase – DLR carried 121.8 million passengers in 2018–19, up 1.8% on the previous year. There are no extensions imminently planned, the proposed Dagenham Dock line now planned to be constructed as a London Overground link to Barking Riverside. However, in late 2016 Mayor of London Sadiq Khan instructed TfL to take forward a new extension of DLR from Gallions Reach (on the Beckton branch) to Thamesmead. This would cross the Thames and help support new housing developments to the north of Woolwich.

In 2015 DLR began an industry consultation on a possible new generation of rolling stock to continue to meet the growth in passenger numbers. Following a competitive tender, in June 2019 CAF was selected as preferred bidder to build 43 new fixed-formation 5-car trains for DLR. The trains will enter service from 2023 and will be built at the CAF plants in Spain. 33 trains will be used to replace all 94 of the single B90/B92/B2K units, with the other ten sets being used to increase service frequencies to and from Stratford and to extend the Beckton–Canning Town shuttle to Stratford International. DLR also has the option to order up to a further 34 units within six years. As part of the upgrade Beckton depot will be expanded with a new 4-road maintenance shed, new stabling sidings and a relocated test track.

CLASS B90 — 2-SECTION UNITS

Built: 1991–92 by BN Construction, Bruges, Belgium. Chopper control.
Wheel Arrangement: B-2-B.
Traction Motors: 2 x Brush DC 140 kW (187 hp).
Seats: 52 (+4).
Weight: 36 tonnes.
Dimensions: 28.80 x 2.65 m.
Braking: Rheostatic and regenerative.
Couplers: Scharfenberg.
Maximum Speed: 50 mph.
Doors: Sliding. End doors for staff use.

22
23
24
25
26
27
28
29
30
31
32
33
34
35
36
37
38
39
40
41
42
43
44

CLASS B92 — 2-SECTION UNITS

Built: 1992–95 by BN Construction, Bruges, Belgium. Chopper control.
Wheel Arrangement: B-2-B.
Traction Motors: 2 x Brush DC 140 kW (187 hp).
Seats: 52 (+4).
Weight: 36 tonnes.
Dimensions: 28.80 x 2.65 m.
Braking: Rheostatic and regenerative.
Couplers: Scharfenberg.
Maximum Speed: 50 mph.
Doors: Sliding. End doors for staff use.

45
46
47
48
49
50
51
52
53
54
55
56
57
58
59
60
61
62
63
64
65
66
67
68
69
70
71
72
73
74
75
76
77
78
79
80
81
82
83
84
85
86
87
88
89
90
91

CLASS B2K — 2-SECTION UNITS

Built: 2002–03 by Bombardier Transportation, Bruges, Belgium.
Wheel Arrangement: B-2-B.
Traction Motors: 2 x Brush DC 140 kW (187 hp).
Seats: 52 (+4).
Weight: 36 tonnes.
Dimensions: 28.80 x 2.65 m.
Braking: Rheostatic and regenerative.
Couplers: Scharfenberg.
Maximum Speed: 50 mph.
Doors: Sliding. End doors for staff use.

92
93
94
95
96
97
98
99
01
02
03
04
05
06
07
08
09
10
11
12
13
14
15
16

▲ B90 and B92 cars 39 and 62 undergo maintenance in the principal DLR depot at Beckton on 7 April 2017. **Robert Pritchard**

▼ Clearly showing the elevated nature of the Beckton line, B2007 cars 108, 132 and 112 (nearest camera) leave Prince Regent with a service for Beckton on 7 April 2017. **Robert Pritchard**

CLASS B2007 — 2-SECTION UNITS

Built: 2007–10 by Bombardier Transportation, Bautzen, Germany.
Wheel Arrangement: B-2-B.
Traction Motors: 4 x VEM AC 130 kW (174 hp).
Seats: 52 (+4).
Weight: 38.2 tonnes.
Dimensions: 28.80 x 2.65 m.
Braking: Rheostatic and regenerative.
Couplers: Scharfenberg.
Maximum Speed: 50 mph.
Doors: Sliding. End doors for staff use.

101	112	123	134	145
102	113	124	135	146
103	114	125	136	147
104	115	126	137	148
105	116	127	138	149
106	117	128	139	150
107	118	129	140	151
108	119	130	141	152
109	120	131	142	153
110	121	132	143	154
111	122	133	144	155

CT30 DIESEL MECHANICAL CRANE TROLLEY

Built: 1986 by Wickham. Hertfordshire. Used for shunting, or for maintenance of overhead or lineside structures. Works No. 11622.
Engine: Ford Sabre 6 cylinder.
Wheel Arrangement: Bo.
Dimensions: 6.75 x 2.63 m.
Weight: 8.25 tonnes.
Maximum Speed: 25 mph.

992

BATTERY ELECTRIC WORKS LOCOMOTIVE WITH CRANE

Built: 1991 by RFS Industries, Kilnhurst, rebuilt 2014 by Hunslet Engine Company Ltd, Leeds.
Battery: Oldham Compton 750 V DC (fully charged).
Traction Motors: 1 x 150 kW.
Wheel Arrangement: B.
Dimensions: 7.55 x 2.50 m.
Weight: 22 tonnes.
Maximum Speed: 12 mph.

993 KYLIE

DIESEL-HYDRAULIC SHUNTING LOCOMOTIVE

Built: 1979 by GEC Traction Ltd, Vulcan Foundry, Newton-le-Willows. One of three locomotives originally built for Shotton steelworks.
Wheel Arrangement: B.
Engine: Gardner 6LXB 165 bhp.
Dimensions: 6.48 x 2.59 m.
Weight: 20 tonnes.
Maximum Speed: 16 mph.

994 KEVIN KEANEY

▲ A comparison of front ends at Poplar depot on 7 April 2017, with (from left to right), cars 73, 07, 64 and 154 stabled awaiting their next duties. **Robert Pritchard**

▼ Interior of B92 car No. 50 showing the mix of facing and longitudinal seating. **Robert Pritchard**

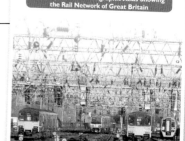

3.3. EDINBURGH TRAMS

Network: 8.5 miles. **Lines:** 1. **Depots:** 1 (Gogar). **First opened:** 2014.
www.edinburghtrams.com **System:** 750 V DC overhead.

The newest tramway in the UK is in Edinburgh, the capital of Scotland with a population of around 500000. It finally opened on 31 May 2014, after several years of delays. The current 8½ mile route is around half of what had been planned to be built, hence the large fleet of 27 trams. The whole project, managed by the since disbanded Transport Initiatives Edinburgh (TIE) was dogged by delays, escalating costs, construction problems and contractual disputes. This saw the scheme, on which construction started in 2008, significantly scaled back to an Edinburgh city centre–Edinburgh Airport tramway, serving the suburbs of Balgreen and Saughton and the business parks around Edinburgh Gateway and Edinburgh Park. A tram stop was also provided for the Murrayfield rugby stadium. In March 2019 City of Edinburgh Council authorised an extension to Newhaven.

While in the suburbs the tramway is mainly segregated, in the city centre there is street running via Haymarket station, Princes Street and St Andrews Square to the terminus at York Place, close to the city's bus station. Unfortunately, there is not a stop ideally placed for Waverley station, passengers being advised to use the St Andrews Square stop and walk down the hill.

The tramway is operated by Edinburgh Trams Ltd, a publicly owned company that works in partnership with Lothian Buses, as part of the Transport for Edinburgh Group.

Work on constructing the 2.8-mile extension eastwards from York Place to Newhaven, with seven stops at Picardy Place (replacing the current York Place terminus), McDonald Road, Balfour Street, Foot of the Walk, The Shore, Port of Leith, Ocean Terminal and Newhaven, will start in late 2019, with completion scheduled for late 2022 and the first passengers due to be carried in February 2023. The fleet of 27 trams is already adequate to serve this planned extension.

A fleet of 27 7-section articulated CAF Urbos cars mounted on four bogies were ordered for the tramway. The trams are of a similar design to those now used by Midland Metro. At 42 m these are the longest trams to operate in the UK and the interior specification is high, with leather seats and large luggage racks for airport passengers. All trams offer free wi-fi. As well as the 78 seats there is room for 170 standing passengers. All trams were delivered in 2011–12, well ahead of the opening of the tramway. The tram numbers follow on the sequence used for the old Edinburgh tramway, which closed in 1956. The livery is similar to that used on Lothian buses – white, burgundy and grey, with black window surrounds. All trams now have advertisments on the upper bodyside and some across the windows, but none are in full advertising livery.

Services operate from around 05.30 to midnight, although the last departure from the Airport is much earlier than that at 22.48. All services operate the full length of the line during the day, although some early morning and late evening trams terminate at Gyle Centre before running to or from the depot. Daytime frequency was improved at the start of 2017, and again in June 2018. Trams now operate approximately every 7 minutes during the daytime, and every 3 minutes at peak times, with an end-to-end journey time of 37 minutes.

The off-peak service all week requires 12 trams in traffic from the fleet of 27. The diagramming for the peak hours is complex, with three extra trams generally used at peak times. This nominally means that 15 trams should be seen in service during a whole weekday, but such is the size of the fleet that different trams may appear on the evening peak diagrams compared to those used in the mornings.

The depot and control centre are located at Gogar, near the Gogarburn stop, and given the large tram fleet a number of vehicles can normally be seen stabled in the yard. There are connections on both sides of the depot, so trams can leave empty for the short run to the Airport or to Gyle Centre to start in traffic. In terms of observing the trams, anywhere on the route should see all those in traffic, although Princes Street – Edinburgh's main shopping street – is an interesting place to watch trams mingling with buses, including special buses offering tours of the city. The Transport for Edinburgh mobile phone app should also be mentioned: as well as live running information this gives details of which trams are in traffic at any given time.

Tickets must be purchased before boarding the tram from one of the ticket machines located on the tram stop platforms. Single tickets start from £1.70, with a day ticket valid on trams and Lothian buses costing £4 for the City zone (all trams apart from Ingliston–Airport), or £9

EDINBURGH TRAMS

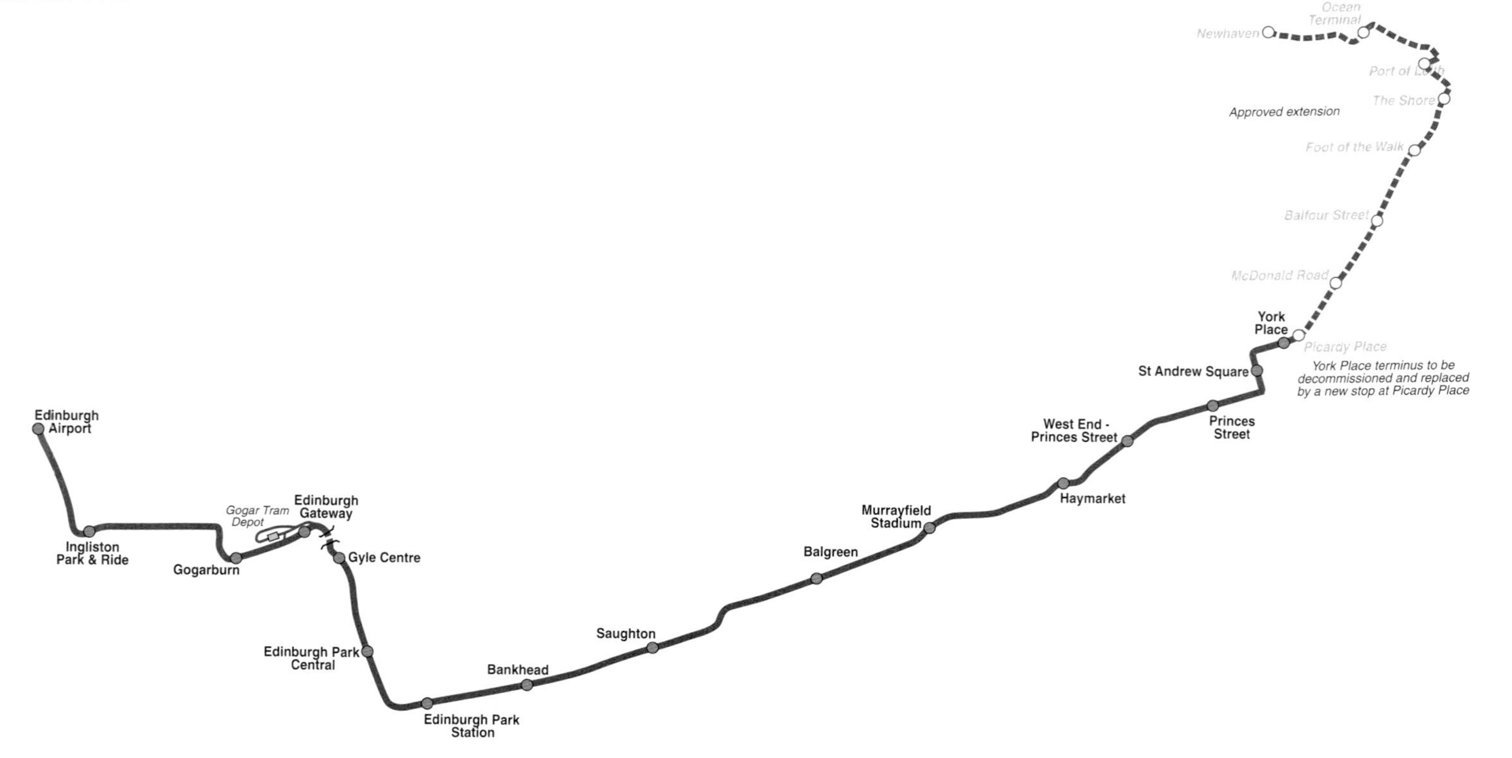

Edinburgh Airport
Ingliston Park & Ride
Gogarburn
Gogar Tram Depot
Edinburgh Gateway
Gyle Centre
Edinburgh Park Central
Edinburgh Park Station
Bankhead
Saughton
Balgreen
Murrayfield Stadium
Haymarket
West End - Princes Street
Princes Street
St Andrew Square
York Place
Picardy Place
York Place terminus to be decommissioned and replaced by a new stop at Picardy Place
McDonald Road
Balfour Street
Foot of the Walk
The Shore
Port of Leith
Ocean Terminal
Newhaven
Approved extension

for the Airport zone, covering all of the tram route. The excessive £5 additional charge for ¾ mile of track between Ingliston and the Airport (walkable in 15 minutes) has received quite a lot of criticism, especially when Airlink buses complete the same journey from the centre of Edinburgh for a cheaper fare than the trams.

Passenger numbers to date have been encouraging, and in 2018 there were 7.3 million passenger journeys, a year-on-year growth of 10%. Passenger numbers are expected to rise to 16 million in the first full year that the Newhaven extension is opened. As well as the Newhaven extension, in 2019 City of Edinburgh Council published bold plans for the expansion of the network, including an extension over the North Bridge to Edinburgh Royal Infirmary and a loop connecting Haymarket with the University of Edinburgh.

CAF URBOS 3 — 7-SECTION TRAMS

Built: 2009–11 by CAF, Irun, Spain.
Wheel Arrangement: Bo-Bo-2-Bo.
Seats: 78.
Dimensions: 42.80 x 2.65 m.
Couplers: Albert.
Doors: Sliding plug.
Traction Motors: 12 x CAF of 80 kW (107 hp).
Weight: 56.25 tonnes.
Braking: Regenerative & electro hydraulic.
Maximum Speed: 50 mph.

251
252
253
254
255
256
257
258
259
260
261
262
263
264
265
266
267
268
269
270
271
272
273
274
275
276
277

▲ 273 arrives at Ingliston Park & Ride with a service for Edinburgh Airport on 6 June 2019.
Ian Lothian

▲ Carrying colourful adverts for the Parabola property investment company, 276 approaches the current terminus at York Place on 7 April 2019. **Robert Pritchard**

▼ 277, heading to the Airport, and 271, heading to York Place pass at Edinburgh Park on 16 September 2018. **Tony Christie**

▲ A general view over the depot at Gogar on 16 September 2018, showing trams stabled outside.
Tony Christie

▼ The plush interior of tram 251 showing on the left one of the dedicated luggage spaces for Airport passengers.
Robert Pritchard

3.4. GLASGOW SUBWAY

Network: 6.5 miles. **Lines:** 1 (circular). **Depots:** 1 (Broomloan). **First opened:** 1896.
www.spt.co.uk/subway **System:** 600 V DC third rail.

This circular 4ft (1219mm) gauge underground line is the smallest metro system in the UK, running for 6½ miles underneath Glasgow city centre and the city's west end. Although not the capital, Glasgow is Scotland's largest city, with a population of around 600 000. Owned and operated by Strathclyde Partnership for Transport the system has 15 stations, eight of which are to the north of the River Clyde and seven to the south. The steepest gradient is 1 in 20, between Govan and Partick. The entire passenger railway is underground, contained in twin-bore tunnels, allowing for clockwise operation on the "Outer" circle and anti-clockwise operation on the "Inner" circle. There are crossovers between the Inner and Outer circles between Govan and Ibrox stations.

The system is the world's third oldest underground railway after London and Budapest, and will celebrate its 125th anniversary in 2021. When opened it was the world's only cable-operated underground railway, driven by a stationary steam boiler. It changed to electric operation in 1935 using a 600 V DC third rail system. Even after electrification the original wooden-bodied vehicles were still used, being converted to electric units. They remained in traffic for some 80 years, with the Subway closing for major modernisation of the stations, tracks and rolling stock in May 1977. A number of the older cars are preserved and these are listed here. Four (2, 16, 20 and 57) have also been converted to works wagons and are still in use on the Subway.

Metro-Cammell was awarded the contract to build 33 new single power cars, which would normally operate in pairs as 2-car units. They used GEC traction motors. In 1992 these were joined by eight intermediate trailer cars, whose bodyshells were constructed at Hunslet TPL in Leeds. Final assembly took place at Hunslet Barclay, Kilmarnock. This brought the fleet size to 41 vehicles, allowing 3-car operation. All seating is longitudinal.

On reopening in April 1980 the system was nicknamed the "Clockwork Orange", inspired by the orange livery carried by the trains. The current, third, livery carried by the existing rolling stock is orange, grey and white and was introduced in 2011. Various cars have carried advertising liveries at different times.

Today all trains are formed of 3-cars: this is either two power cars sandwiching one of the trailer cars or three power cars. Trains run in semi-automatic mode in passenger service, meaning that speed and stopping are controlled by an Automatic Train Operation (ATO) system.

It takes 24 minutes to complete the loop in either direction. Trains run every 4 minutes at peak times, so 12 trains are needed for the peak service – this means 36 of the available 40 vehicles (car 122 has been stood down as a source of spares and subsequently scrapped) are required. At off-peak times during the day trains run every 6 minutes, and then every 8 minutes after 19.00 and on Sundays (the system only operates between 10.00 and 18.00 on Sundays).

Both the depot and Subway control centre are at Broomloan, near Govan. At the time of the Subway modernisation in the late 1970s, access tracks to the depot were built: previously trains had to be lifted from the main tracks into the maintenance shed using a crane. Today all trains return to the depot overnight. The depot access tracks and the depot itself are the only part of the system that is above ground. The depot complex includes a test track, mainly used for testing trains after maintenance. A new test track is currently under construction as part of the introduction of the new fleet of trains. Also based here is the engineering locomotive fleet.

In March 2016 it was announced that Stadler had won the tender to supply the next generation of trains for the Subway, as part of a £288 million upgrade, with Italian company Ansaldo responsible for resignalling the system. There will be 17 new 4-car driverless trains, which will feature through gangways and a new forward view for passengers to enjoy. The first completed train was exhibited to the media at the InnoTrans trade show in Berlin in September 2018. Formed of two identical half trains coupled together, each set is 39.24 m long, slightly longer than the existing 3-car sets. The two centre cars are shorter, and articulated. TSA of Austria is suppling the traction motors for the new trains.

The first of the new trains, which will be numbered in the 301–317 series, was delivered in May 2019, and the fleet is planned to enter service in 2020–21, initially manually driven using a temporary driver's cab, which will later be removed: from 2023 it is planned to introduce "Unattended Train Operation" (UTO) with no driver. This will be the first example of full UTO in

GLASGOW SUBWAY

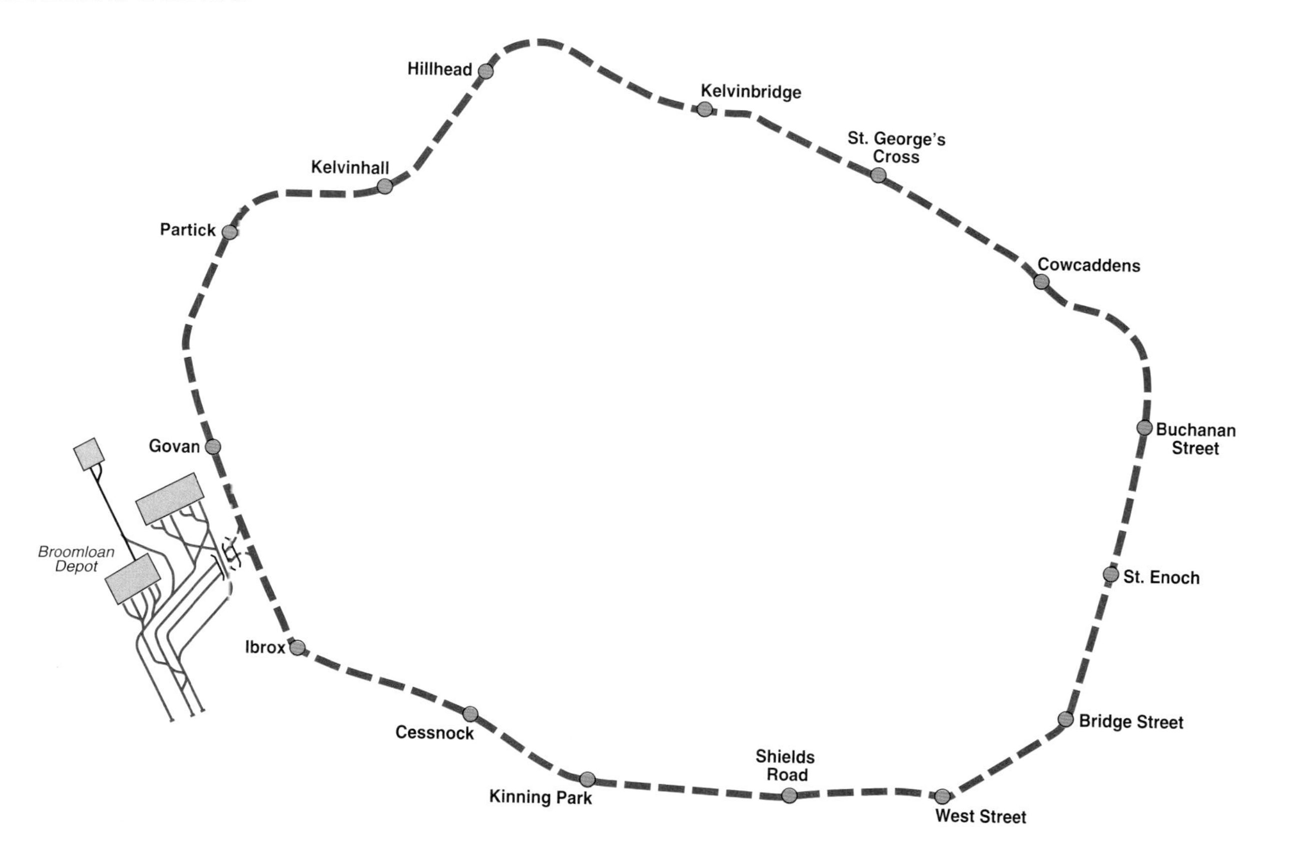

Britain, as the Docklands Light Railway (see separate section) does still have Passenger Service Agents on board its trains. Full UTO is used on driverless metros such as in Lille, on Paris Metro Lines 1 and 14 and in Copenhagen, Denmark. Signalling will be the Communications Based Train Control (CBTC) supplied by Ansaldo that is also used in Copenhagen, Milan and Stockholm. Resignalling will enable the current 4-minute headway to be reduced to just 1½ minutes and it is planned that 16 out of the 17 new units will be diagrammed at peak times. The new units have a mainly white livery, with black window surrounds and orange doors.

Comprehensive refurbishment of each of the stations started in 2011 and had been completed at all but four stations by spring 2019. In addition all 28 of the system's escalators have been replaced. The 123-year-old system and the design of the smaller Subway station platforms means it is impracticable to make all Subway stations wheelchair accessible, although where this is possible this has been done. Lifts have been provided at St Enoch (the closest station to Glasgow Central) and Govan. As part of the upgrade for new rolling stock "half-height" platform edge screen doors will be installed at stations before UTO operation takes place.

All stations are staffed and have ticket gates which check your ticket on entry and exit. Tickets can be purchased from ticket offices or from machines. A single ticket costs £1.75 (£1.55 with a smartcard) in 2019, and an all-day ticket is £4.20 (£3 with smartcard). For £7.20 the good-value Glasgow Roundabout Ticket includes the Subway and also suburban rail services around Glasgow, or for £12.70 the Strathclyde Daytripper is valid on the Subway, rail services, most buses and some ferries across the wider Strathclyde region. Roundabout and Daytripper holders will need to present their ticket at the Subway ticket office to obtain a gate pass, however.

Various expansion plans have been discussed over the years but none have been progressed further so the network today is largely the same as it was when it opened in 1896. 13.1 million passengers were carried on the Subway in 2018–19, the same as for the previous year, but up from the 11.1 recorded in 2016–17 when services were disrupted by rebuilding works.

DRIVING POWER CARS

Built: 1977–79 by Metro-Cammell, Birmingham. Refurbished 1993–95 by ABB Derby.
Wheel Arrangement: Bo-Bo.
Traction Motors: 4 x GEC G312AZ of 35.6 kW (48 hp).
Seats: 36.
Weight: 19.6 tonnes.
Dimensions: 12.81 m x 2.34 m.
Braking: Disc.
Couplers: Wedgelock.
Maximum Speed: 33.5 mph.
Doors: Sliding.
Non-Standard Liveries: 101 – Original Glasgow Subway livery (maroon & cream).
130 – New Glasgow Subway livery (white & orange).

101 **0**
102
103
104
105
106
107
108
109
110
111
112
113
114
115
116
117
118
119
120
121
123
124
125
126
127
128
129
130 **0**
131
132
133

INTERMEDIATE BOGIE TRAILERS

Built: 1992 by Hunslet TPL, Leeds/Hunslet Barclay, Kilmarnock.
Wheel Arrangement: 2-2.
Traction Motors: –
Seats: 40.
Weight: 17.2 tonnes.
Dimensions: 12.70 m x 2.34 m.
Braking: Disc.
Couplers: Wedgelock.
Maximum Speed: 33.5 mph.
Doors: Sliding.

201
202
203
204
205
206
207
208

NEW GENERATION SUBWAY EMUs

Built: 2018–20 by Stadler, Bussnang, Switzerland. New build Subway trains.
Wheel Arrangement: Bo-Bo+2+2+Bo-Bo.
Formation: DM–T–T–DM.
Seats: 96(+6) (with cabs); 104(+6) (without cabs).
Dimensions: 13 + 7 + 7 + 13 m.
Couplers: Scharfenberg.
Doors: Sliding.
Traction Motors: TSA of 520 kW (total).
Weight: 55.4 tonnes.
Braking: Disc and regenerative.
Maximum Speed: 36 mph.

301
302
303
304
305
306
307
308
309
310
311
312
313
314
315
316
317

BATTERY ELECTRIC WORKS LOCOMOTIVES

Built: 1977 by Clayton Equipment, Hatton.
Battery: Battery pack of 94 lead acid cells (188V).
Wheel Arrangement: Bo.
Traction Motors: 2 x GEC G312AZ of 35.6 kW (48 hp) series wound.
Dimensions: 4.70 m x 2.10 m.
Couplers: Wedgelock.
Weight: 14.8 tonnes.
Maximum Speed: 15 mph.

L2 (S) LOBEY DOSSER
L3 RANK BAJIN

BATTERY ELECTRIC WORKS LOCOMOTIVES

Built: 2010 by Clayton Equipment, Hatton.
Battery: Battery pack of 100 lead acid cells (200V).
Wheel Arrangement: Bo.
Traction Motors: 2 x 56 kW (75hp) series wound.
Dimensions: 6.61 m x 2.10 m.
Couplers: Wedgelock.
Weight: 15 tonnes.
Maximum Speed: 15 mph.

L6
L7

BATTERY ELECTRIC WORKS LOCOMOTIVES

Two new locomotives delivered in 2017 for shunting the new Stadler trains.
Built: 2016–17 by Clayton Equipment, Burton-upon-Trent.
Battery: Battery pack of 159 cells (320V).
Wheel Arrangement: Bo.
Traction Motors: 2 x SR series of 52 kW (70 hp).
Dimensions: 5.41 m x 2.08 m.
Couplers: Scharfenberg.
Weight: 16 tonnes.
Maximum Speed: 12.5 mph.

Unnumbered (Works No. B4624A)
Unnumbered (Works No. B4624B)

3.4.1. GLASGOW SUBWAY PRESERVED STOCK

A number of the original Glasgow Subway vehicles have been preserved, as listed below.

GLASGOW DISTRICT SUBWAY COMPANY CARS

Built: 1896–1901 by Oldbury Railway Carriage & Wagon Company, Oldbury as Gripper cars, converted to motor cars 1935, or 1897–98 by Hurst Nelston & Company, Motherwell as Trailer cars.

Gripper	12.42 x	m	tons	42 seats.
Trailer	7.62* x	m	tons	24 seats. * Later extended to 12 m.

1	Gripper	Glasgow Museum Resources Centre, Nitshill	Oldbury 1896
4†	Gripper	Glasgow Riverside Museum	Oldbury 1896
7	Gripper	*Location unknown*	Oldbury 1896
39	Trailer	Glasgow Riverside Museum	Hurst Nelson 1898
41†	Trailer	Glasgow Museum Resources Centre, Nitshill	Hurst Nelson 1898
55	Gripper	Bo'ness & Kinneil Railway	Oldbury 1901

† Only part of the original vehicles survives.

▲ Carrying the new mainly white Subway livery, 130 leads 129 and 132 into St Enoch with an Outer Circle service on 15 April 2019. 130 is the only one of the old vehicles to receive this livery. **Robert Pritchard**

▼ Interior of Driving Power Car 130. **Robert Pritchard**

▲ All of the public lines of the Subway are below ground, but part of the link line from the depot is above ground. On 5 March 2019 an unidentified driving car, trailer 203 and driving car 107, has just left the depot and is heading onto the system to start an evening peak diagram. **Marc Nicol**

▼ The first of the new Stadler Subway trains (301), on the right, is seen after arrival at Broomloan depot on 4 May 2019, alongside one of the old sets, led by car 104. **Courtesy SPT**

3.5. GREATER MANCHESTER METROLINK

Network: 59 miles. **Lines**: 8. **Depots**: 2 (Queens Road and Trafford). **First opened**: 1992.
www.metrolink.co.uk **System**: 750 V DC overhead.

Metrolink was the first modern tramway system in the UK when it opened in 1992, combining street running in the city centre with longer distance running over former BR lines. Conversion of the Bury–Victoria line to a tramway and the replacement of obsolete EMUs on this route was first proposed as part of a 1982 study, which suggested a six-route light rail system. When the system opened it had 26 trams and operated from Bury to Altrincham with a street section through the centre of Manchester and a spur to Piccadilly station (crucially linking Victoria and Piccadilly stations). Since then, in comparison with most of the other UK tramways, Metrolink has expanded considerably, the number of extensions opening in the last ten years being testament to the dedication of Greater Manchester PTE (now Transport for Greater Manchester (TfGM)) and its faith in the continued growth of the Metrolink system.

Manchester is a city of just over 500000 population, although the Metropolitan area of Greater Manchester also includes many other large towns, such as Bury, Salford, Oldham, Rochdale, Stockport and Bolton, some places now served by the enlarged Metrolink network.

Metrolink is owned by Transport for Greater Manchester, the transport delivery arm for the Greater Manchester Combined Authority which is made up of the ten Greater Manchester councils. TfGM is responsible for investment in Metrolink, its transport services and facilities, but sub-contracts the day-to-day operation of the tramway. Metrolink is operated by KeolisAmey, who took over from RATP Dev in July 2017. KeolisAmey (the same partnership that operates Docklands Light Railway in London) has a 10-year contract to operate Metrolink until 2027. The new operator pledged to recruit 300 more staff and to increase staff presence on the network.

After the initial Bury/Altrincham–Piccadilly route, the first extension was a line from Cornbrook to Eccles which opened in 1999–2000. This route features quite a lot of street-running. After a gap of 10 years the next round of extensions followed in quick succession, almost tripling the length of the system in 5 years. First was a short spur off the Eccles line to MediaCityUK, which opened in September 2010. The first part of the South Manchester Line to Chorlton and St Werburgh's Road opened in summer 2011 and this was further extended (along the former Manchester South District Railway and Midland Railway line) to East Didsbury in 2013.

It had long been an aspiration to convert the Oldham loop to a tramway, to provide a more frequent service and take people into the centre of Manchester. The national rail Oldham loop line closed for rebuilding in October 2009, reopening from Manchester to Oldham Mumps in summer 2012. The original alignment here only lasted a year and a half, being replaced by a new alignment through Oldham town centre which opened at the start of 2014. Meanwhile the top end of the loop line to Rochdale reopened as a tramway in late 2012/early 2013. A further extension into Rochdale town centre followed in spring 2014. Both the Oldham and Rochdale lines feature some street running.

East of Piccadilly the East Manchester Line reached Droylsden in February 2013 and this was followed by Droylsden–Ashton-under-Lyne in October 2013, on a route that features more longer sections of street running. In late 2014 St Werburgh's Road–Manchester Airport opened, offering a mix of segregated and street running and connecting the communities of Baguley and Wythenshawe. The Second City Crossing, a short but important section of track in the city centre, opened between 2015 and 2017. The original city crossing via Mosley Street was by now getting very congested and in order to cross more trams from one side of the city to the other a second route was needed. The route chosen was from St Peter's Square, which was rebuilt with four platforms, along Cross Street to Exchange Square and then to Victoria. The first part of the Second City Crossing, known locally as "2CC" from Victoria to Exchange Square, opened in late 2015 and then the remainder to St Peter's Square in February 2017. The Second City Crossing is used by trams from East Didsbury to Rochdale or Shaw & Crompton and from Manchester Airport to Victoria.

The opening dates of the various lines are summarised below:

6 April 1992: Bury–Victoria.
27 April 1992: Victoria–Deansgate-Castlefield.
15 June 1992: Deansgate–Altrincham.
20 July 1992: Market Street–Piccadilly.
6 December 1999: Cornbrook–Broadway.

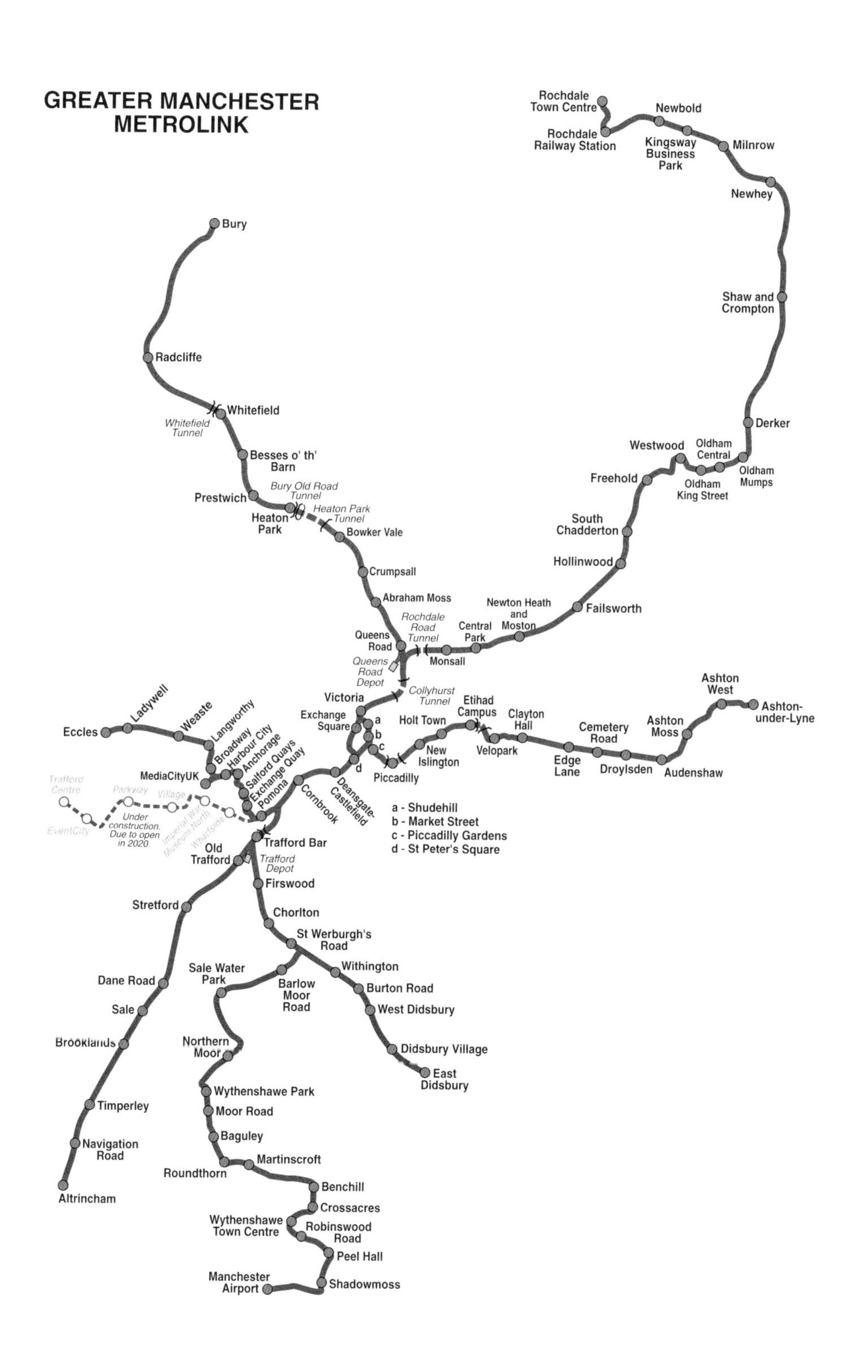
GREATER MANCHESTER
METROLINK
Rochdale Town Centre
Rochdale Railway Station
Newbold
Kingsway Business Park
Milnrow
Newhey
Shaw and Crompton
Derker
Oldham Mumps
Oldham Central
Oldham King Street
Westwood
Freehold
South Chadderton
Hollinwood
Failsworth
Newton Heath and Moston
Central Park
Monsall
Rochdale Road Tunnel
Bury
Radcliffe
Whitefield
Whitefield Tunnel
Besses o' th' Barn
Prestwich
Bury Old Road Tunnel
Heaton Park
Heaton Park Tunnel
Bowker Vale
Crumpsall
Abraham Moss
Queens Road
Queens Road Depot
Collyhurst Tunnel
Victoria
Exchange Square
a
b
c
d
Holt Town
Etihad Campus
New Islington
Piccadilly
Velopark
Clayton Hall
Edge Lane
Cemetery Road
Droylsden
Audenshaw
Ashton Moss
Ashton West
Ashton-under-Lyne
Eccles
Ladywell
Weaste
Langworthy
Broadway
Harbour City
Anchorage
Salford Quays
Exchange Quay
Pomona
MediaCityUK
Trafford Centre
EventCity
Parkway
Village
Imperial War Museum North
Wharfside
Under construction. Due to open in 2020.
Cornbrook
Deansgate-Castlefield
a - Shudehill
b - Market Street
c - Piccadilly Gardens
d - St Peter's Square
Trafford Bar
Old Trafford
Trafford Depot
Firswood
Stretford
Chorlton
St Werburgh's Road
Withington
Burton Road
West Didsbury
Didsbury Village
East Didsbury
Dane Road
Sale
Brooklands
Timperley
Navigation Road
Altrincham
Sale Water Park
Barlow Moor Road
Northern Moor
Wythenshawe Park
Moor Road
Baguley
Martinscroft
Roundthorn
Benchill
Crossacres
Wythenshawe Town Centre
Robinswood Road
Peel Hall
Manchester Airport
Shadowmoss

21 July 2000: Broadway–Eccles.
20 September 2010: Harbour City–MediaCityUK.
7 July 2011: Trafford Bar–St Werburgh's Road.
13 June 2012: Victoria–Oldham Mumps.
16 December 2012: Oldham Mumps–Shaw & Crompton.
8 February 2013: Piccadilly–Droylsden.
28 February 2013: Shaw & Crompton–Rochdale Railway Station.
23 May 2013: St Werburgh's Road–East Didsbury.
9 October 2013: Droylsden–Ashton-under-Lyne.
27 January 2014: Freehold–Derker via Oldham Central.
31 March 2014: Rochdale Railway Station–Rochdale Town Centre.
3 November 2014: St Werburgh's Road–Manchester Airport.
6 December 2015: Victoria–Exchange Square.
26 February 2017: Exchange Square–St Peters Square (Second City Crossing).

It was decided at an early stage that Manchester would be a high-floor network, because of the existing high platforms on the Bury and Altrincham lines inherited from BR. Therefore on the street running sections, high-floor platforms were constructed.

Despite now having 93 stops, almost 60 miles of track, eight different routes, and two different routes across Manchester city centre, Metrolink doesn't use route numbers, letters or colours on its trams – only on the route map where letters (and different colours to indicate the routes) are used. All routes run to a 12 minute frequency at both peak and off-peak times on Mondays–Saturdays (until around 20.00). This gives a service every 6 minutes on large parts of the system – for example on the original lines to Bury and Altrincham and also to Ashton-under-Lyne, Harbour City, St Werburgh's Road and Shaw & Crompton.

Changes since the last edition of this book have seen the extension of route D from Piccadilly to Ashton-under-Lyne, but the truncation of line B at Piccadilly instead of Etihad Campus. Line F has been extended from Deansgate-Castlefield to Victoria, via the Second City Crossing. This most recent route extension took place in January 2019. The route letters used on the map and the journey time of the different routes are as follows:

A: Altrincham–Market Street–Bury: journey time 57 minutes;
B: Altrincham–Piccadilly: journey time 32 minutes;
C: Bury–Piccadilly: journey time 33 minutes;
D: MediaCityUK–Piccadilly–Ashton-under-Lyne: journey time 54 minutes;
E: Eccles–Piccadilly–Ashton-under-Lyne: journey time 63 minutes;
F: Manchester Airport–Exchange Square–Victoria: journey time 58 minutes;
G: East Didsbury–Exchange Square–Rochdale Town Centre: journey time 1h18;
H: East Didsbury–Exchange Square–Shaw & Crompton: journey time 1h02.

The current service requires 103 of the 120 M5000 trams. Metrolink is the only UK tram network that uses pairs of trams in normal service. On Mondays–Saturdays all services on route A (Altrincham–Bury, 11 diagrams) are diagrammed for pairs of trams, as are two diagrams each on routes A and C and four on routes G/H (these two routes interwork, for example a diagram will run East Didsbury–Rochdale–East Didsbury–Shaw & Crompton–East Didsbury etc). All diagrams on Sundays are normally for single trams.

Most routes operate from around 05.30 to midnight. In 2016 very early morning services were introduced on the Airport route, aimed at airport workers or those passengers with early flights. The first service runs empty from Trafford depot to Firswood where it leaves at 03.00, arriving at the Airport at 03.39. Trams then run every 20 minutes from Deansgate-Castlefield until 06.00, when the normal 12-minute frequency starts. Additional trams operate in connection with football matches at Old Trafford (Manchester United) or the Etihad Stadium (Manchester City).

On Sundays routes A, D and H do not run and route E from Ashton-under-Lyne to Eccles runs via MediaCityUK, using the north side of the triangle near Harbour City. The other five routes operate to the standard 12-minute frequency on Sundays.

The original fleet consisted of 26 articulated two-section T68 trams supplied by Firema of Italy, which were numbered 1001–1026. These were supplemented by six similar vehicles (2001–2006), built in 1999 for the Eccles extension. By 2007 more trams were needed, initially to alleviate overcrowding and also to serve planned new extensions. The "Flexity Swift" M5000 model from Bombardier was selected, closely based on the K5000 high-floor cars supplied for the Cologne Stadtbahn light rail system in Germany. The M5000s offered a similar capacity to a T68, although with around 30 fewer seats, but 146 standing spaces. The first M5000 (numbered 3001) arrived in July 2009 and entered traffic in December of that year.

▲ 3081 passes through a quiet Rochdale town centre on Good Friday, 19 April 2019, with a service for East Didsbury. **Robert Pritchard**

▼ 3103 approaches Manchester Airport on 12 November 2017. **Robert Pritchard**

The last T68s were withdrawn from normal service in April 2014. The majority were scrapped but 1003 found further use with the Greater Manchester Fire Service for training purposes and four of the class were moved to Long Marston for UKTram. A further four are still stored at Metrolink depots, two of which are reserved for the Heaton Park Tramway.

The M5000s have been ordered in a number of different batches between 2007 and 2018. The first 12 were ordered in 2007 – eight to alleviate overcrowding on the existing system and four for the new MediaCityUK extension. This was followed in June 2008 by an order for another 28 for the extensions to Oldham/Rochdale and St Werburgh's Road. Another eight were ordered in March 2010 for the East Didsbury and Ashton extensions, and this was followed by another 14 in July 2010 for the Airport line and the Oldham and Rochdale town centre lines.

As thoughts turned to replacement of the ageing T68s, 12 more M5000s were ordered in September 2011 to replace the 12 T68s in the worst condition. In July 2012 there was an order for another 20 M5000s after the decision was taken to withdraw all the T68s. In January 2014 ten more were ordered for the forthcoming Trafford Park extension (although not due to open until 2020) and this was followed by an order for 12 in July 2014 and a final four in September 2014, both to enable more double trams to operate.

The most recent order was for a further 27 cars, to be delivered in 2020–21. This order was placed in July 2018 and will enable all services on the Bury and Altrincham routes to be formed of pairs of trams, as well as enabling additional double-trams on the East Didsbury–Shaw/ Rochdale routes plus some on the Ashton-under-Lyne route.

The M5000s are assembled at the Bombardier plant in Vienna, but the sub-assemblies and primary parts are manufactured in Bautzen, Germany, with the underframes coming from Česká Lípa in the Czech Republic and electrical equipment supplied by Vossloh Kiepe. The M5000s are double-ended articulated trams with three bogies, the two outer bogies being powered. From car 3075 upwards the seating layout was modified, with eight more seats (60 instead of 52).

Although outwardly the same, the M5000s are operationally split into two fleets: 3001–3060 are fitted with Automatic Tram Stop (ATS) and Vehicle Recognition System (VRS) and can be used on all routes. 3061–3120 are not fitted with these systems and are only used on the routes to Rochdale, Manchester Airport, East Didsbury and Ashton-under-Lyne.

When first opened with T68s the Metrolink livery was white, grey and blue. When the M5000s were delivered the livery was changed to silver and yellow.

The M5000s are maintained at both depots. The original depot and workshops is located at Queens Road, adjacent to the Bury line. A second depot opened in 2010 to maintain the growing fleet, this is located at Trafford, close to Old Trafford station.

Since the opening of the Second City Crossing, the busiest stop on the network is St Peter's Square, where all trams except those operating route C between Piccadilly and Bury can be viewed. The Piccadilly Gardens triangle is also a good place to observe the trams, but those services using the Second City Crossing cannot be seen here.

Tickets must be purchased from the ticket machines at stops before boarding the tram. Single and return tickets are available, with a city centre zone single costing £1.40. A Peak Day Travelcard valid on trams only is £7, an Off-Peak Travelcard costing £4.80 and one for a weekend (valid after 18.00 Fridays) costing £6.60. There are various other tickets available also covering buses and trains within Greater Manchester, and also the GM Wayfarer, which for £14 allows off-peak travel on trains, trams and buses throughout Greater Manchester and also parts of Cheshire, Derbyshire, Lancashire, Staffordshire and the Peak District.

Not surprisingly, considering it is by far the largest tramway network in the UK, Metrolink in 2018–19 carried 43.7 million passengers according to DfT figures, an increase of 6.1% on the previous year despite no new extensions opening during this time.

Work started in 2017 on a 3½ mile extension from Pomona to Trafford Centre, which will have five intermediate stops and is planned to open in late 2020. It is anticipated that trams from Trafford Centre will operate to Crumpsall via the Second City Crossing. Ambitious plans for further expansion of the network, including a number of tram-train style routes, were unveiled in early 2019 as part of the Greater Manchester Transport Strategy 2040 Draft Delivery Plan 2020–25. Options being looked at include the extension of the Airport line to Davenport Green and the Airport Terminal 2, plus routes from Ashton to Stalybridge, East Didsbury–Stockport and Trafford Centre to Port Salford. Tram-train routes could include Rochdale–Heywood–Bury, Stockport–Manchester Airport, Colnbrook–Manchester Airport via Timperley, plus a number of current national rail lines, such as to Marple and Wigan via Atherton.

T68 1000 SERIES — 2-SECTION TRAMS

Built: 1991–92 by Firema, Italy. Chopper control.
Wheel Arrangement: Bo-2-Bo.
Seats: 82 (+4).
Dimensions: 29.00 x 2.65 m.
Couplers: Scharfenberg.
Doors: Sliding.
Traction Motors: 4 x GEC of 130 kW (174 hp).
Weight: 45 tonnes.
Braking: Regenerative, disc & emergency track.
Maximum Speed: 50 mph.
Livery: White, dark grey & blue with light blue doors.

1007, 1020 and 1023 are stored at Trafford depot.
1016, 1022, 1024 and 1026 are stored at Long Marston for use in UKTram development work.

1007	(S)	EAST LANCASHIRE RAILWAY
1016	(S)	
1020	(S)	
1022	(S)	POPPY APPEAL
1023	(S)	
1024	(S)	
1026	(S)	

T68 2000 SERIES — 2-SECTION TRAMS

Built: 1999 by Ansaldo, Italy. Chopper control.
Wheel Arrangement: Bo-2-Bo.
Seats: 82 (+4).
Dimensions: 29.00 x 2.65 m.
Couplers: Scharfenberg.
Doors: Sliding.
Traction Motors: 4 x GEC of 130 kW (174 hp).
Weight: 45 tonnes.
Braking: Regenerative, disc & magnetic track.
Maximum Speed: 50 mph.
Livery: White, dark grey & blue with light blue doors.

2001 is stored at Trafford depot.

2001 (S)

▲ St Peter's Square is the best stop to observe trams, with four platforms and 35 movements in each direction per hour on weekdays. On 19 April 2019 3081 arrives with a service for Rochdale via the Second City Crossing, with 3073 to the right with a service to Ashton-under-Lyne. **Robert Pritchard**

▲ 3093 (one of four trams in a red advertising livery for Vodaphone) and 3099 are seen between Shaw & Crompton and Derker with a Rochdale–East Didsbury service on 19 April 2019. **Robert Pritchard**

▼ Interior of M5000 tram 3079 showing the wheelchair space in the foreground. **Robert Pritchard**

3000 SERIES FLEXITY SWIFT 2-SECTION TRAMS

Built: 2009–20 by Bombardier, Vienna, Austria.
Wheel Arrangement: Bo-2-Bo.
Traction Motors: 4 x Vossloh Kiepe 3-phase asynchronous of 120 kW (161 hp).
Seats: 52 (3001–3074) or 60 (3075–3147).
Dimensions: 28.40 x 2.65 m.
Couplers: Scharfenberg.
Doors: Sliding.
Weight: 39.7 tonnes.
Braking: Regenerative, disc & magnetic track.
Maximum Speed: 50 mph.
Livery: Silver & yellow.
Advertising Liveries: 3017, 3047, 3093 and 3100 – Vodaphone (red).
3066 – Clean air Greater Manchester (light blue).
3081 – IKEA (white).
3094 – Olivia Burton London (white).
3099 – Gymshark (black/yellow).
3116 – Cheshire Oaks Designer Outlet (grey).

3004, 3005, 3014, 3016, 3017, 3021, 3032, 3051 and 3055 are fitted with ice-breaking pantographs.

3121–3147 are on order and are due to arrive between February 2020 and June 2021.

3001
3002
3003
3004
3005
3006
3007
3008
3009
3010
3011
3012
3013
3014
3015
3016
3017 **AL**
3018
3019
3020
3021
3022
3023
3024
3025
3026
3027
3028
3029
3030
3031
3032
3033
3034
3035
3036
3037
3038
3039
3040
3041
3042
3043
3044
3045
3046
3047 **AL**
3048
3049
3050
3051
3052
3053
3054
3055
3056
3057
3058
3059
3060
3061
3062
3063
3064
3065
3066 **AL**
3067
3068
3069
3070
3071
3072
3073
3074
3075
3076
3077
3078
3079
3080
3081 **AL**
3082
3083
3084
3085
3086
3087
3088
3089
3090
3091
3092
3093 **AL**
3094 **AL**
3095
3096
3097
3098
3099 **AL**
3100 **AL**
3101
3102
3103
3104
3105
3106
3107
3108
3109
3110
3111
3112
3113
3114
3115
3116 **AL**
3117
3118
3119
3120
3121
3122
3123
3124
3125
3126
3127
3128
3129
3130
3131
3132
3133
3134
3135
3136
3137
3138
3139
3140
3141
3142
3143
3144
3145
3146
3147

Names:

3009 50th Anniversary of Coronation Street 1960–2010
3020 LANCASHIRE FUSILIER
3022 Spirit of MCR
3098 Gracie Fields

SPECIAL PURPOSE VEHICLE

Purpose-built 4-wheel diesel-hydraulic multi-purpose locomotive (with crane) for shunting, engineering or tram recovery work. Normally based at Queens Road depot.

Built: 1991 by RFS Industries, Kilnhurst.
Engine: Caterpillar 3306 PCT of 170 kW.
Dimensions: 8.20 x 2.50 m.
Weight: 22 tonnes.
Wheel Arrangement: Bo.
Transmission: Hydraulic. Rockwell T280.
Couplers: Scharfenberg.
Maximum Speed: 25 mph.

1027

3.6. LONDON TRAMLINK

Network: 17.5 miles. **Lines**: 4. **Depots**: 1 (Therapia Lane). **First opened**: 2000.
www.tfl.gov.uk/trams **System**: 750 V DC overhead.

The London Tramlink system, initially called "Croydon Tramlink", runs through central Croydon (one of the southern Greater London boroughs) via a one-way loop, with lines radiating out to Wimbledon, New Addington, Beckenham Junction and Elmers End. The system opened on 10 May 2000, returning trams to the streets of London for the first time in almost 50 years (the New Addington branch opened first, followed by the other lines later in May 2000). Tramlink is operated by Tram Operations Ltd, a subsidiary of the major train and bus operator First Group, on behalf of Transport for London.

Tramlink grew out of a transport study carried out by London Transport and BR in 1986 covering all of Greater London, plus a need to provide better rail or tram services in the area and link previously separate routes. While most of the network serves areas which already had rail connections, New Addington had never before had any kind of rail service, and this is served by a mostly new alignment purpose-built for trams, plus part of the former Elmers End–Sanderstead rail route (closed in 1983), including the 500 m tunnel just beyond Sandilands. Unlike most other UK street-running tramways, little of the on-street sections shares road space with private cars.

Much of the rest of the system uses former railway lines: the Wimbledon line and part of the eastern route which covers most of the old Elmers End–Addiscombe branch. Both the Wimbledon and Addiscombe branches closed in 1997 to allow for conversion to the tramway. These are combined with some street running through a large central Croydon city loop, which as part of the current timetable is used by trams from New Addington (Line 3) running in a clockwise direction. In the centre of Croydon eastbound trams on Lines 2 and 4 run via Tamworth Road, West Croydon station and Wellesley Road while westbound trams use George Street. The hub of the operation is East Croydon main line station, where all services can be seen.

The service pattern was changed in February 2018 to take advantage of a new second platform at Wimbledon and additional double track sections on the Wimbledon line, with Line 1 from Elmers End to Croydon being withdrawn and Line 2 from Beckenham Junction being altered to run to Wimbledon. This gave Wimbledon a 5-minute frequency service, but reduced the Elmers End frequency from every 7–8 minutes to every 10 minutes. The peak service is roughly the same as the off-peak service, with 20 trams an hour passing East Croydon in each direction, reduced slightly from 22 as the two longer Wimbledon routes now require trams running higher mileages through to Wimbledon. Elmers End services. The three routes are as follows:

Line 2: Beckenham Junction–East Croydon–Wimbledon (every 10 minutes and every 15 minutes evenings and Sundays);
Line 3: New Addington–West Croydon (every 7/8 minutes and every 15 minutes late evenings);
Line 4*: Elmers End–East Croydon–Wimbledon (every 10 minutes, every 15 minutes late evenings and Sundays).

* Elmers End services which do not run to Wimbledon but only to Therapia Lane or the Croydon loop in the evenings run as Line 1.

Both the Tramlink depot and control centre are located at Therapia Lane, which is on the Wimbledon line. Tramlink operates for close to 22 hours a day, in terms of the last trams arriving at the depot and the first leaving the next morning. Compared to some other systems early morning and late evening positioning trips operate as normal services, rather than empty stock. For example, the working timetable shows the first departure from Therapia Lane depot at 04.14 to New Addington, with the last arrival back at Therapia Lane being at 01.43 (01.10 from Beckenham Junction). The longest journey on the network is on Line 2, which takes 51 minutes end-to-end.

When the system opened in 2000 it was operated by a fleet of 24 Bombardier 3-section "Flexity Swift" six-axle CR4000 trams, closely based on the K4000 trams built for the Cologne low-floor network in Germany. These are mostly low-floor, although there is a short high-floor section at the outer end over the motor bogies. The unusual number series (2530 onwards) carries on from where the old London tram fleet left off.

Although there have been no extensions to the original system, passenger growth has been such that a new fleet of trams was required. The first of six Stadler Variobahn trams (the first three of these had been built ahead of schedule for Bergen, Norway but were diverted to

Wimbledon
Dundonald Road
Merton Park
Morden Road
Phipps Bridge
Belgrave Walk
Mitcham
Mitcham Junction
Beddington Lane
Therapia Lane Depot
Therapia Lane
Ampere Way
Waddon Marsh
Wandle Park
West Croydon
Centrale
Reeves Corner
Church Street
George Street
Wellesley Road
East Croydon
Lebanon Road
Sandilands
Sandilands Tunnels
Addiscombe
Blackhorse Lane
Woodside
Arena
Elmers End
Harrington Road
Birkbeck
Avenue Road
Beckenham Road
Beckenham Junction
Lloyd Park
Coombe Lane
Gravel Hill
Addington Village
Fieldway
King Henry's Drive
New Addington

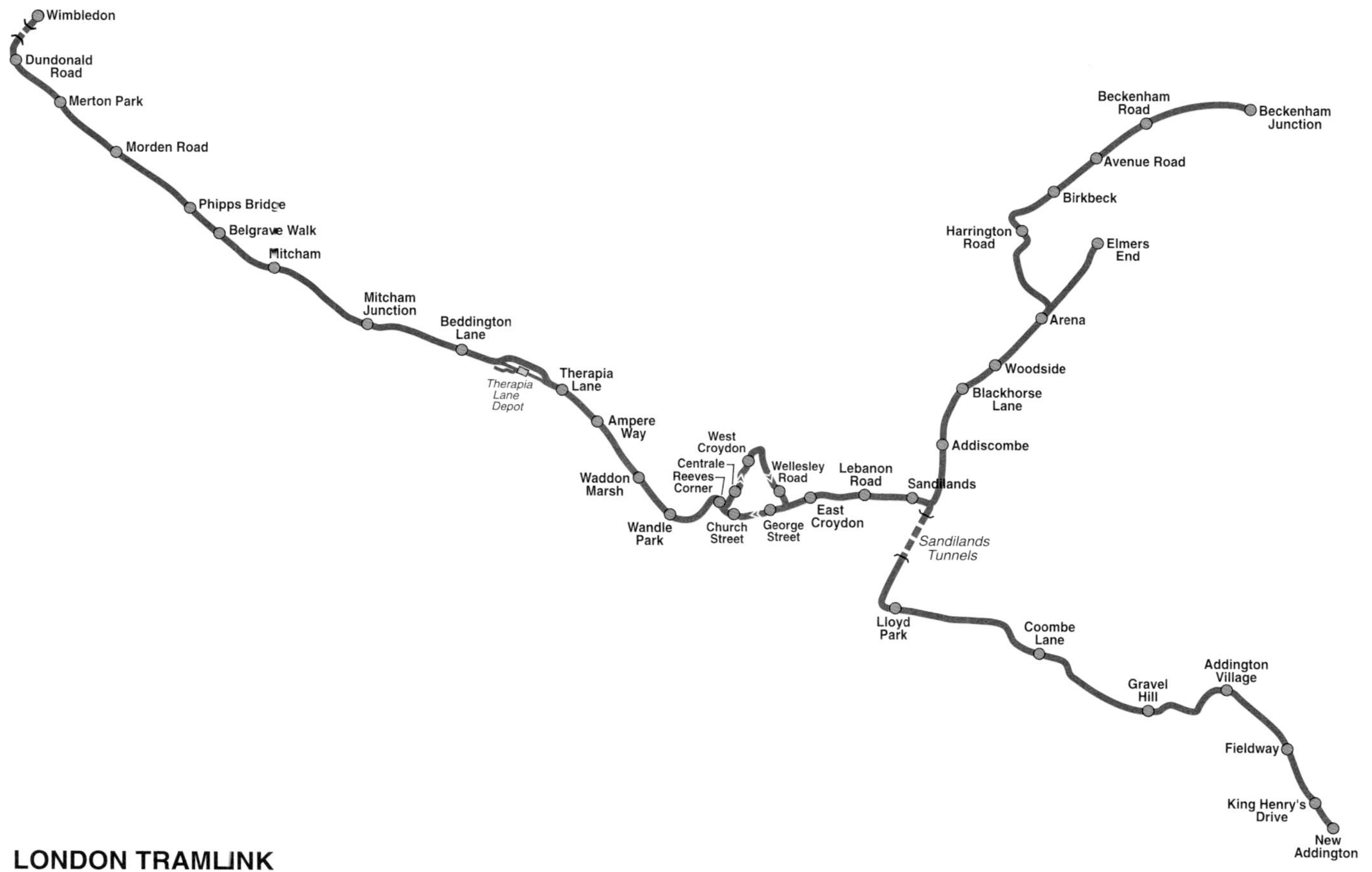

LONDON TRAMLINK

Croydon), which are 5-section, fully low-floor trams and slightly longer than the original trams, entered traffic in spring 2012. Smaller orders followed for another four (2560–63) which arrived in 2015 and then a final two (2564–65) which arrived in 2016. The 12 Stadler trams are used across the whole system on all routes alongside the original fleet, with a preference for using them on the longest route (Line 2).

All trams always operate singly. The current weekday daytime timetable requires 30 trams in operation from the serviceable fleet of 35 (not including the long-time the out-of-traffic 2551): 12 trams on Line 2, eight on Line 3 and ten on Line 4.

On 9 November 2016 London Tramlink suffered a serious accident, when tram 2551 overturned on the tight curve at Sandilands, the tram having taken the curve at more than three times the permitted speed. Seven people lost their lives in the accident, which is the worst in modern UK tramway history. Tram 2551 is now stored at an RAIB site at Aldershot.

The original livery was red and white, based on that of the old London trams. In 2008 this was changed to light grey and lime green, with a blue solebar, the lime green colour being used to denote Tramlink on the TfL zonal Travelcard map. Various advertisement or promotional liveries have been carried by some trams at different times.

London Tramlink forms an integral part of the TfL network, with interchange available at several places to heavy rail or LUL (at Wimbledon). The whole of the network is within Travelcard Zone 4. A single fare with a contactless bankcard or Oystercard smartcard is £1.50. Cards must be "touched in" before boarding the tram (but not "touched out"). A Zones 1–4 or 1–6 Travelcard is valid on all tram services. See www.oyster-rail.org.uk/trams-and-buses/ and www.oyster-rail.org.uk/wimbledon/ for details of how to "sign in" and "sign out" at Wimbledon and Elmers End.

28.7 million passengers used Tramlink in 2018–19, a slight decrease, although Tramlink is still the second busiest UK tramway (as opposed to a Metro) network after Manchester Metrolink. TfL has been consulting on a proposed extension of the network to Sutton, with three possible routes proposed, one of which would involving converting the heavy rail Sutton loop to light rail operation. Although there have been concerns over TfL's spending cuts, it is hoped that construction of a Sutton extension could start in 2022 with opening following three years later. Plans for a second turning loop in Croydon (the Dingwall Road loop) have been put on hold following the withdrawal of funding from the Westfield shopping centre.

BOMBARDIER FLEXITY SWIFT 3-SECTION TRAMS

Built: 1998–99 by Bombardier, Vienna, Austria.
Wheel Arrangement: Bo-2-Bo.
Seats: 70.
Dimensions: 30.10 x 2.65 m.
Couplers: Scharfenberg.
Doors: Sliding plug.
Traction Motors: 4 x 120 kW (161 hp).
Weight: 36.3tonnes.
Braking: Disc, regenerative and magnetic track.
Maximum Speed: 50mph.

2530	2535	2540	2545	2550
2531	2536	2541	2546	2551 (S)
2532	2537	2542	2547	2552
2533	2538	2543	2548	2553
2534	2539	2544	2549	

Name:

2535 STEPHEN PARASCANDOLO 1980–2007

STADLER VARIOBAHN 5-SECTION TRAMS

Built: 2011–16 by Stadler, Berlin, Germany.
Wheel Arrangement: Bo-2-Bo.
Seats: 74.
Dimensions: 32.40 x 2.65 m.
Couplers: Albert.
Doors: Sliding plug.
Traction Motors: 8 x 45kW (60 hp).
Weight: 41.5tonnes.
Braking: Disc, regenerative and magnetic track.
Maximum Speed: 50mph.
Advertising Livery: 2554 – Love Croydon (purple & blue).

2554 **AL**	2557	2560	2562	2564
2555	2558	2561	2563	2565
2556	2559			

▲ Springtime in New Addington: on 5 April 2019 Bombardier Flexity Swift 2545 approaches King Henry's Drive with a service for New Addington. **Robert Pritchard**

▼ 2550 is seen between Gravel Hill and Addington Village with a New Addington Service on 5 April 2019. **Robert Pritchard**

▲ Stadler car 2562 heads away from the camera on the single track central Croydon loop near Centrale with a service for Elmers End on 9 August 2018. **Robert Pritchard**

▼ Interior of Flexity Swift car 2553. **Robert Pritchard**

3.7. NOTTINGHAM EXPRESS TRANSIT

Network: 20 miles. **Lines:** 2. **Depots:** 1 (Wilkinson Street). **First opened:** 2004.
www.thetram.net **System:** 750 V DC overhead.

Nottingham is a city of around 320000. The Nottingham Express Transit light rail system was first conceived in 1988 as a joint venture between Nottingham City Council and Nottinghamshire County Council. Approved in 1998, construction began in 2001 and the tramway opened on 9 March 2004, with a line from Station Street running for around 8 miles north to Hucknall, plus a 1 mile spur to Phoenix Park, where there is a large Park & Ride car park. The first part of this route involves around 3 miles of street running in Nottingham, including through old Market Square and a bi-directional single track section through the Hyson Green area. The northern part of the route runs mainly parallel with the Nottingham–Mansfield–Worksop railway line. North of Bulwell both the tramway and railway line are single track, with crossing loops at tramway stops. At Hucknall there is a railway/tramway interchange station.

On 25 August 2015 the system more than doubled in length, with the opening of two extensions south and west from Station Street. At the same time the Station Street stop was moved onto a new viaduct which now carries trams over Nottingham station. The two new lines diverge almost immediately south of the station, one route running to Clifton South via Wilford and the other to Toton Lane (Chilwell) via Beeston. Both routes feature long sections of on-street running and part of the Clifton South branch follows the alignment of the old Great Central line from Nottingham Victoria towards Loughborough. The Toton Lane line serves the large ng2 business park, Queen's Medical Centre and Nottingham University, as well as the residential districts of Beeston and Chilwell.

Trams run on two unnumbered routes, the longest being Hucknall–Toton Lane (green on the NET maps), which takes 63 minutes and the other being Phoenix Park–Clifton South (purple on the maps) which takes 46 minutes. The system now has 51 stops.

Nottingham Express Transit was originally operated by a joint consortium of Nottingham City Council and Transdev, but is now operated by Nottingham Trams Ltd, a consortium of Keolis (80%) and Wellglade (20%).

Trams operate from around 06.00 to midnight. The weekday off-peak services is every 10 minutes on both routes, this gives a 5-minute frequency on the Wilkinson Street–Nottingham Station section. Peak services on weekdays operate between the hours of 07.00–10.00 and 15.00–19.00 when the frequency is every 7 minutes on both routes, providing a service every 3–4 minutes on the core centre section. The Saturday service is unusual in that between 10.00 and 19.00 it operates to the weekday peak frequency; every 7 minutes on both routes. At other times it is every 10 or 15 minutes. On Sundays it is every 15 minutes in the early morning and evening and every 10 minutes during the day.

32 trams out of the overall fleet of 37 are required for the peak service. Off peak the Hucknall–Toton Lane service requires 15 trams, plus three additionals at peak times. The Phoenix Park–Clifton South route requires ten trams off-peak, and an additional four at peak times.

The original fleet consisted of 15 Bombardier built Incentro AT6/5 trams (built to an Adtranz design). These were supplemented by 22 new Alstom Citadis 302 trams, of a similar design to those built for Dublin. These were delivered between September 2013 and August 2014 for the new extensions. Both the Incentro and Citadis trams are bi-directional, 100% low-floor trams with five articulated sections riding on two motor bogies (at the outer ends) and one trailer bogie. In the Citadis trams, as well as the 58 seats (plus ten tip-ups) there is space for 144 standing passengers. The Incentro trams have space for 129 standing. The first of the new Alstom trams entered service in July 2014 on the existing system. As both types of tram have a similar capacity they are used interchangeably across both routes. Refurbishment of the original Incentro trams started in spring 2019 with car 203 and including a revised external livery.

Livery is silver and dark green, with black window surrounds. NET has named all of its trams after personalities or famous sportsmen or women from the local area (although some names are now only carried on one side or at one end on the older cars). Two trams receive new names every year. Tram 220 is renamed each year to celebrate Nottingham's Nurse of the Year, whilst 222 is the designated "Community Hero" tram, and receives a new name every year on the anniversary of the tramways reopening (in March) to celebrate local community heroes.

The depot and workshops are located at Wilkinson Street and were expanded with additional stabling sidings when the new Alstom trams arrived. There are connections from the depot for trams to both head north (towards Hucknall) or south (towards the city). Alstom is contracted to maintain both the original Bombardier trams and its own Citadis trams.

A single ticket between any two stops on the network is £2.30. A day ticket offering unlimited travel is £4. Tickets valid on trams and Trent Barton buses are also available. Nottingham trams originally used conductors on all trams who sold tickets, but when the 2015 extensions opened this was changed to a system of roving ticket inspectors. Now all tickets must be purchased from ticket machines at stops before boarding the tram.

In terms of observing the trams, the Nottingham Station stop is good, with all services passing through. Old Market Square stop in the city centre is also a pleasant place to observe, although it can be tricky for photographs. The Wilkinson Street stop outside the depot has the added interest of trams running on and off the depot before and after the morning/evening peaks.

Passenger numbers more than doubled since the 2015 extensions opened, and continue to grow steadily. The 2018–19 passenger figures show 18.8 million passengers used NET, an increase of 5.7% on the previous year.

A number of other extensions have been discussed, in particular extending the Toton Lane line to terminate at the planned new High Speed 2 parkway station at Toton. A feasibility study is in place into a possible 3-mile extension from Phoenix Park to Kimberley. More ambitious proposals have included a tramway linking Nottingham and Derby.

BOMBARDIER INCENTRO — 5-SECTION TRAMS

Built: 2002–03 by Bombardier, Derby Litchurch Lane Works.
Wheel Arrangement: Bo-2-Bo. **Traction Motors:** 8 x 45 kW (65 hp) wheelmotors.
Seats: 54 (+4). **Weight:** 36.7 tonnes.
Dimensions: 33.00 x 2.40 m. **Doors:** Sliding plug.
Braking: Disc, regenerative and magnetic track for emergency use.
Couplers: Not equipped. **Maximum Speed:** 50 mph.
Non-Standard/Advertising Liveries: 206 – jet2.com (red & blue).
209 – Big event (lime green).
210 – Just Eat (various)
211 – Deliveroo (green, purple & blue).

No.	Livery	Name
201		Torvill and Dean
202		DH Lawrence
203		Bendigo Thompson
204		Erica Beardsmore
205		Lord Byron
206	**AL**	Angela Alcock
207		Mavis Worthington
208		Dinah Minton
209	**AL**	Sidney Standard
210	**AL**	Sir Jesse Boot
211	**AL**	Robin Hood
212		William Booth
213		Mary Potter
214		Dennis McCarthy MBE
215		Brian Clough OBE

ALSTOM CITADIS 302 — 5-SECTION TRAMS

Built: 2013–14 by Alstom, Barcelona, Spain.
Wheel Arrangement: Bo-2-Bo. **Traction Motors:** 4 x 120 kW (161 hp).
Seats: 58 (+10). **Weight:** 40.8 tonnes.
Dimensions: 32.00 x 2.40 m. **Doors:** Sliding plug.
Braking: Disc, regenerative and magnetic track for emergency use.
Couplers: Not equipped. **Maximum Speed:** 50 mph.

216	Dame Laura Knight
217	Carl Froch MBE
218	Jim Taylor
219	Alan Sillitoe
220	Unisa Avanzado
221	Stephen Lowe
222	David S Stewart OBE
223	Colin Slater MBE
224	Vicky McClure
225	Doug Scott CBE
226	Jimmy Sirrel & Jack Wheeler
227	Sir Peter Mansfield
228	Local Armed Forces Heroes
229	Viv Anderson MBE
230	George Green
231	Rebecca Adlington OBE
232	William Ivory
233	Ada Lovelace
234	George Africanus
235	David Clarke
236	Sat Bains
237	Stuart Broad MBE

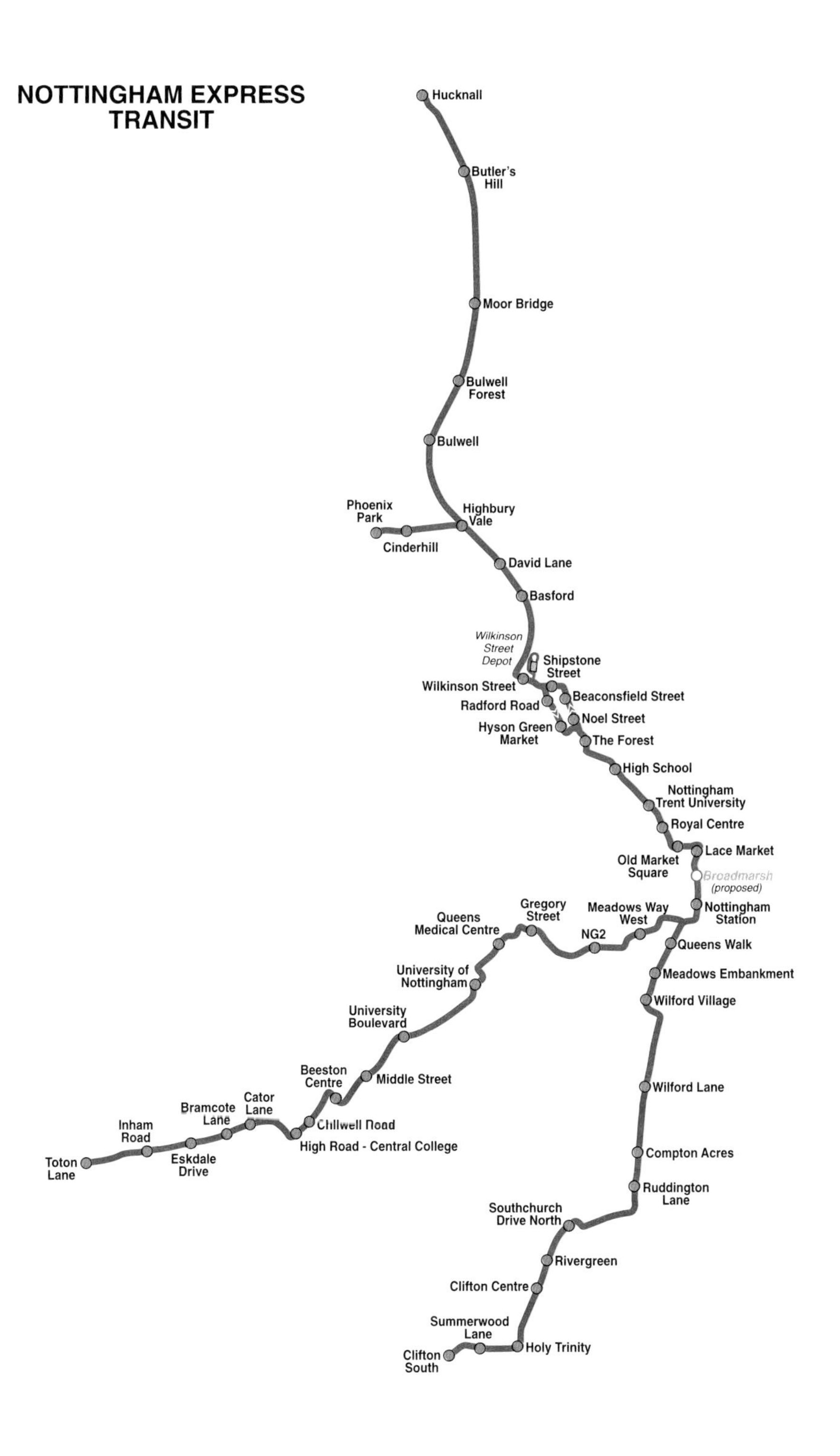
NOTTINGHAM EXPRESS TRANSIT
Hucknall
Butler's Hill
Moor Bridge
Bulwell Forest
Bulwell
Phoenix Park
Cinderhill
Highbury Vale
David Lane
Basford
Wilkinson Street Depot
Shipstone Street
Wilkinson Street
Beaconsfield Street
Radford Road
Noel Street
Hyson Green Market
The Forest
High School
Nottingham Trent University
Royal Centre
Lace Market
Old Market Square
Broadmarsh (proposed)
Nottingham Station
Gregory Street
Meadows Way West
Queens Medical Centre
NG2
Queens Walk
Meadows Embankment
University of Nottingham
Wilford Village
University Boulevard
Beeston Centre
Middle Street
Wilford Lane
Cator Lane
Bramcote Lane
Chilwell Road
Inham Road
High Road - Central College
Toton Lane
Eskdale Drive
Compton Acres
Ruddington Lane
Southchurch Drive North
Rivergreen
Clifton Centre
Summerwood Lane
Holy Trinity
Clifton South

▲ One of the original Bombardier cars, 208 "Dinah Minton", arrives at the Queens Walk stop with a Phoenix Park–Clifton service on 9 April 2019. **Robert Pritchard**

▼ A comparison of front ends of the old and new trams at The Forest on 30 June 2018, on the left Bombardier car 204 "Erica Beardsmore" leaves for Toton Lane, whilst Alstom car 232 "William Ivory" arrives heading to Phoenix Park. **Robert Pritchard**

▲ Alstom Citadis tram 219 "Alan Sillitoe" has just left the Nottingham Station stop with a Hucknall–Toton Lane service on 9 April 2019. **Robert Pritchard**

▼ Interior of one of the Bombardier cars, 203, showing one of the areas of 2+1 seating. **Robert Pritchard**

3.8. SOUTH YORKSHIRE SUPERTRAM

Network: 22 miles. **Lines:** 4. **Depots:** 1 (Nunnery). **First opened:** 1994.
www.supertram.com **System:** 750 V DC overhead.

Sheffield is a city of around 500000. The South Yorkshire Supertram system opened in stages during 1994 and 1995, having been first conceived during feasibility studies in the 1980s, with tramway construction beginning in 1991. The first extension to the original Supertram network came in 2018 when a trial tram-train service started operating to Rotherham.

The tram network has three lines radiating from Sheffield city centre. These run to Halfway in the south-east with a spur from Gleadless Townend to Herdings Park, to Middlewood in the north with a spur from Hillsborough to Malin Bridge and to Meadowhall Interchange in the north-east, adjacent to the large shopping complex. The tram-train service leaves the Meadowhall line just after Meadowhall South station for Rotherham Central before terminating at Parkgate. Opening dates for the different sections were as follows:

21 March 1994: Fitzalan Square–Meadowhall;
22 August 1994: Fitzalan Square–Spring Lane;
5 December 1994: Spring Lane–Gleadless Townend;
17–18 February 1995: Fitzalan Square–Shalesmoor;
27 March 1995: Gleadless Townend–Halfway;
3 April 1995: Gleadless Townend–Herdings Park;
23 October 1995: Shalesmoor–Middlewood/Malin Bridge.
25 October 2018: Meadowhall South–Rotherham Parkgate.

Supertram was the first low-floor tram system in Britain when in opened in 1994, two years after the high-floor Manchester system. It had a mix of on-street and segregated running (54% of track is embedded or street-running rail). The Meadowhall branch is mostly segregated with longer sections of the Hillsborough and Halfway routes having street running. There are now 50 stops on the system, including Rotherham Central and Parkgate.

The system was originally operated by South Yorkshire Supertram Ltd (SYSL), a subsidiary of South Yorkshire Passenger Transport Executive (SYPTE). SYSL was sold to bus and train operator Stagecoach in 1997, and this company has operated the trams since then, branded as "Stagecoach Supertram". Stagecoach is also responsible for the maintenance of the fleet at the Nunnery depot, which is adjacent to the Nunnery Square stop. Stagecoach has an agreement to operate the current Supertram concession until March 2024.

The original fleet of 25 trams (at first numbered 01–25) were obtained from Siemens-Duewag in Germany, and are widely recognised as some of the most comfortable and successful trams in the UK. The cars are owned by South Yorkshire Light Rail Ltd, a subsidiary of SYPTE. They are all-steel, double-ended in three articulated sections with eight powered axles: because of severe gradients on the hills in Sheffield (up to 1 in 10) all axles are powered. The outer sections are low floor (with a high-floor seating section immediately behind each cab) and the centre section, on which the pantograph is located, is high floor. As well as the 86 seats the trams can take 162 standing passengers.

Four of the original trams are not in their original A+B+C formations following accidents. Trams 02 and 11 (as they were then numbered) were damaged in separate accidents in 1995. 02 returned to traffic as two sections of 02 and one of 11 and 11 with two sections of 11 and one of 02. In 2015 cars 118 and 120 collided with each other and one section of 118 was formed with two from 120 to return "120" to traffic. All four trams 102, 111, 118 and 120 remain permanently misformed.

In 2012 the Department for Transport chose Sheffield as the pilot for a tram-train project, which would see tram-trains used on both the existing network and on a new route to the nearby large town of Rotherham. Originally due to be completed by 2016 delays to the track works put the expected start date back several times. It eventually commenced on 25 October 2018, operating for a 2-year trial period, fully funded by the DfT. After the trial is finished it is hoped it can continue until at least the end of the current Stagecoach concession in 2024. Unlike the other three routes, the tram-train route does not have a route specific colour, but tram-trains display "TT" alongside their destination and on route maps the tram-train is shown in black.

The first of the seven new tram-trains, built by Vossloh (now Stadler) in Spain was delivered in late 2015, with the remainder following in 2016. The first of these were introduced onto the

SOUTH YORKSHIRE SUPERTRAM

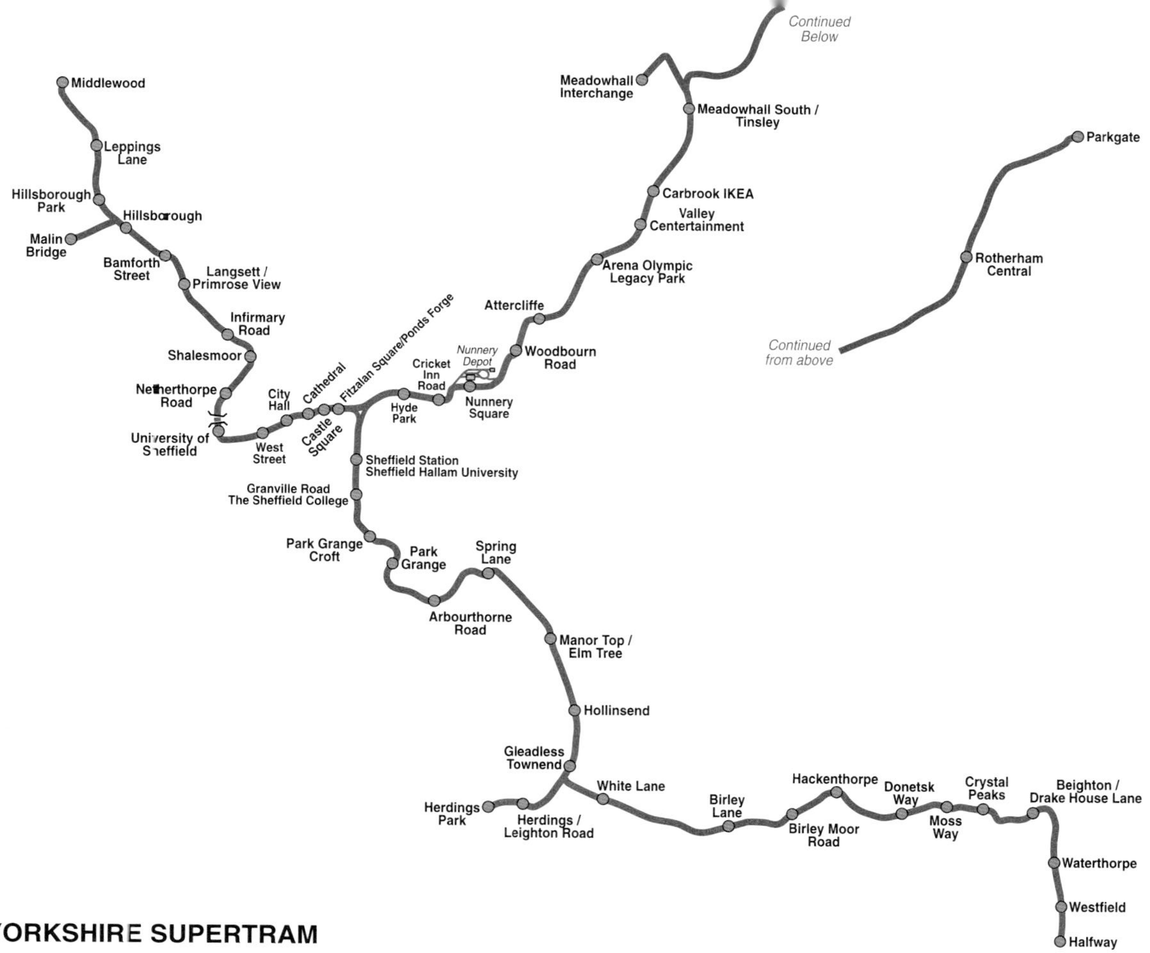

Middlewood
Leppings Lane
Hillsborough Park
Hillsborough
Malin Bridge
Bamforth Street
Langsett / Primrose View
Infirmary Road
Shalesmoor
Netherthorpe Road
University of Sheffield
West Street
City Hall
Cathedral
Castle Square
Fitzalan Square/Ponds Forge
Hyde Park
Cricket Inn Road
Nunnery Depot
Nunnery Square
Woodbourn Road
Attercliffe
Arena Olympic Legacy Park
Valley Centertainment
Carbrook IKEA
Meadowhall South / Tinsley
Meadowhall Interchange
Continued Below
Parkgate
Rotherham Central
Continued from above
Sheffield Station Sheffield Hallam University
Granville Road The Sheffield College
Park Grange Croft
Park Grange
Arbourthorne Road
Spring Lane
Manor Top / Elm Tree
Hollinsend
Gleadless Townend
Herdings / Leighton Road
Herdings Park
White Lane
Birley Lane
Birley Moor Road
Hackenthorpe
Donetsk Way
Moss Way
Crystal Peaks
Beighton / Drake House Lane
Waterthorpe
Westfield
Halfway

existing system in September 2017. These were built as dual voltage vehicles, but the tram-train route is currently electrified at 750 V DC, the same as the rest of the tramway. Should Network Rail electrification extend to the Rotherham Central line in the future the tram-trains could operate on 25 kV AC.

Tram-trains leave the existing system via a new chord near the Meadowhall South/Tinsley stop, joining the current freight only (former Great Central) line from there to Rotherham Central, and then terminating around 1½ miles further north outside the Rotherham Parkgate retail complex, where a dedicated tram-train platform has been built. There are three services per hour, all week, but these are not at regular intervals owing to the timings of passenger trains on the Rotherham Central line. The Sunday service operates only between 09.00 and 18.00.

The tram-trains are branded as Vossloh Citylink cars, and are similar to the original trams, being 3-section with a high-floor centre section (without doors). They also have four bogies, but only three are powered. For operation on Network Rail lines EMU unit numbers 399 201–207 are carried (as well as vehicle numbers in the 999xxx series). Supertram fleet numbers are 201–207. These vehicles are also fitted with safety equipment for operating on the national railway network, such as TPWS, OTMR and GSM-R.

From the outset the seven tram-trams were split into two fleets. Three cars are required for the tram-train service from Cathedral to Parkgate: 201–204 had a slightly different wheel profile for tram-train running – street sections of Supertram have grooved rails that could not be negotiated by the broader wheel profiles that are standard to Network Rail trains, and the wheels of trams are not compatible with NR switch blades. Vehicles with a "tram-train" wheel profile can be used to Rotherham and on city centre track that has been relaid, currently on the existing system between Meadowhall/Meadowhall South and Shalesmoor/Spring Lane. Vehicles with a "tram" wheel profile can be used on all of the traditional tram routes.

Unfortunately, the tram-train service got off to a rather eventful start, as 204 collided with a lorry near Attercliffe on the opening day, vehicle 999004 being badly damaged. The following month 202 was damaged when it collided with a car at the same crossing. This led to 206 being converted to tram-train operation and a hybrid consisting of two vehicles from 202 and one vehicle from 204 being formed. Cars 205 and 207 were left to operate on the rest of the system. The badly damaged vehicles have been returned to Stadler in Spain for repairs, which should be completed by the end of 2019.

The three tram routes are colour coded as Yellow, Blue and Purple. These colours are displayed on the front of trams. Tram-trains are denoted by "TT" and the route is shown in black on Supertram maps. There were some minor changes to the other three routes once the tram-train service started. The normal weekday daytime schedule is now as follows:

Yellow: Middlewood–Cathedral–Meadowhall; Every 12 minutes, journey time 40 minutes;
Blue: Malin Bridge–Cathedral–Halfway; Every 12 minutes, journey time 54–56 minutes;
Purple: Herdings Park–Cathedral. Every 30 minutes, journey time 23 minutes*.
Tram-train: Cathedral–Rotherham Parkgate. Three per hour, but not at regular intervals, journey time 26 minutes.
* Mon–Fri daytime, operates every 20 minutes Cathedral–Herdings Park at peak times.

Trams operate between around 06.00 and midnight on the three coloured routes. Additional peak-hour trips operate on the Yellow and Blue routes. At evenings and on Sundays all three routes operate a 20-minute service meaning that the short Herdings Park branch actually has a better service then, as both the Blue and Purple routes run every 20 minutes, giving a 10 minute core service on the city centre–Gleadless Townend section, and every 20 minutes to Halfway and Herdings Park.

22 trams are required for the peak tram service, with 21 at off-peak times. In addition three tram-trains are required for the Rotherham service at all times. All routes pass through the Cathedral–Fitzalan Square section in the city centre, meaning this is the best place for observing trams. The Cathedral stop can be regarded as the hub of the operation and has the added interest of the tram-trains and Purple route trams reversing. The busy Hillsborough–city centre route also has a tram every 6 minutes as it is covered by both the Yellow and Blue routes. A number of early morning and late evening services use the east side of the Park Square triangle when they are running from the depot to form early morning services or returning to the depot at night. These services can be worked out from the timetable as they are those that start or terminate at Cricket Inn Road and then next call at Sheffield Station.

The trams have carried three different liveries, being delivered in a plain grey livery, this was replaced by a mainly white Stagecoach livery (when they were renumbered from 01–25 to

▲ In an XPO Logistics advertising livery, 116 is seen at the old Sheffield Station stop with a Halfway service (starting here because of engineering works) on 7 April 2019. Behind is 118 in the pink prettylittlething.com advertising livery. **Alan Yearsley**

▼ Carrying the original Sheffield Corporation livery, 120 passes Meadowhall South/Tinsley on 25 October 2018 with an empty working to the depot, as 125 leaves for Meadowhall. **Robert Pritchard**

101–125) and then the present standard Stagecoach blue, red and orange livery, which was first introduced in 2006. Various advertisement liveries have also been carried, and in 2010 tram 120 received a version of the original Sheffield Corporation tram livery of cream and blue to mark 50 years since the closure of the original Sheffield tramway in 1960. This livery is still carried.

Tickets are purchased on board the tram from the conductor. A single ticket costs from £1.80, with an all day Supertram ticket costing £4.20 (this is valid to Rotherham Parkgate for a promotional period), or one for Stagecoach buses and trams in Sheffield costing £4.60. A Sheffield CityWide at £5.00 covers all trams and buses in Sheffield, while a South Yorkshire Connect+ day ticket at £8.50 gives unlimited travel on buses, trams and trains in South Yorkshire.

Sheffield Supertram was used by 11.9 million passengers in 2018–19, a figure that has dropped from the 15 million mark between 2007 and 2012 through a combination of disruption caused by line closures for track relaying works and more competitive ticket pricing by bus operators. Sheffield has the highest percentage of journeys by concessionaries in the country (29.1%).

SIEMENS 3-SECTION TRAMS

Built: 1993–94 by Siemens-Duewag, Düsseldorf, Germany.
Wheel Arrangement: B-B-B-B.
Traction Motors: 4 x monomotor drives of 265 kW (355 hp).
Seats: 80 (+6).
Dimensions: 34.75 x 2.65 m.
Couplers: Albert (emergency use).
Doors: Sliding plug.
Weight: 52 tonnes.
Braking: Regenerative, disc & emergency track.
Maximum Speed: 50 mph.
Non-Standard/Advertising Liveries: 109 – East Midlands Trains (deep pink).
111 and 118 – Prettylittlething.com (pink).
116 – XPO Logistics (white & red).
120 – Original Sheffield Corporation (cream & blue).

101	106	111 **AL**	116 **AL**	121
102	107	112	117	122
103	108	113	118 **AL**	123
104	109 **AL**	114	119	124
105	110	115	120 **0**	125

VOSSLOH/STADLER CITYLINK 3-SECTION TRAM-TRAINS

Built: 2014–15 by Vossloh/Stadler, Valencia, Spain.
Systems: 750 V DC/25 kV AC overhead.
Wheel Arrangement: Bo-2-Bo-Bo.
Seats: 88 (+8).
Dimensions: 37.20 x 2.65 m.
Couplers: Albert (emergency use).
Doors: Sliding plug.
Traction Motors: 6 x VEM of 145 kW (195 hp).
Weight: 64 tonnes.
Braking: Regenerative, disc & emergency track.
Maximum Speed: 60 mph.

Both the full Network Rail EMU numbers 399 201–207 and 3-digit tram numbers are carried. Unlike the other trams, each section carries an individual number.

399 202 and 399 204 are currently misformed. 399 204 has been returned to Spain for repairs after accident damage. It is expected that these trams will stay in these formations.

399 201	999001	999101	999201	
399 202	999002	999102	999204	Theo – The Children's Hospital Charity
399 203	999003	999103	999203	
399 204	999004	999104	999202 (S)	
399 205	999005	999105	999205	
399 206	999006	999106	999206	
399 207	999007	999107	999207	

▲ Vossloh Citylink tram-train 399 203 passes Rotherham United's New York Stadium, just south of Rotherham Central, with a tram-train service for Cathedral on 27 June 2019.
Robert Pritchard

▲ 399 203 leaves the Rotherham Parkgate terminus with a tram-train service to Cathedral on 3 February 2019. **Robert Pritchard**

▼ Interior of the centre section of Citylink tram-train 399 201. **Robert Pritchard**

3.9. TYNE & WEAR METRO

Network: 48 miles. **Lines:** 2. **Depots:** 1 (South Gosforth). **First opened:** 1980.
www.nexus.org.uk/metro **System:** 1500 V DC overhead.

The impressive and efficient Tyne & Wear Metro system covers 48 route miles and can be described as the UK's first modern light rail system. The initial network opened between 1980 and 1984, consisting of a line from South Shields via Gateshead and Newcastle Central to Bank Foot (later extended to Newcastle Airport in 1991) and the North Tyneside loop (over former BR lines) serving Tynemouth and Whitley Bay with a terminus in the city centre at St James. A more recent extension was from Pelaw to Sunderland and South Hylton in 2002, using existing heavy rail infrastructure between Heworth and Sunderland. Opening dates for the various sections were as follows:

11 August 1980: Haymarket–South Gosforth–Whitley Bay–Tynemouth;
10 May 1981: South Gosforth–Bank Foot;
15 November 1981: Haymarket–Heworth;
14 November 1982: St James–Monument–North Shields–Tynemouth;
24 March 1984: Heworth–Pelaw–South Shields;
17 November 1991: Bank Foot–Newcastle Airport;
31 March 2002: Pelaw–Sunderland–South Hylton.

More stations were slowly added in following years to give today's total of 60 stations.

The network runs over a mixture of former heavy rail lines and in newly constructed tunnels under central Newcastle (two new viaducts were also needed). Nine stations are fully underground and Sunderland is the only station where heavy rail and Metro trains share the same platforms.

Operationally there are two routes, coloured Green and Yellow on the official route maps. The Green route runs from the Airport, to the North-West of Newcastle, through the city to Gateshead and south to Sunderland and South Hylton. The journey time for this route is scheduled to take between 1h04 and 1h07 minutes.

The Yellow route is more complex, operating as a loop and branch. Starting from St James in the city centre trains head east to North Shields and then anti-clockwise around the Coast loop via Whitley Bay, through Newcastle and south-east to South Shields. On arrival at South Shields a train will then complete this route in reverse, ie running to Newcastle, clockwise around the Coast loop via Whitley Bay and North Shields to St James. The journey time for a full Yellow route train is scheduled to take 1h24.

The off-peak Monday–Friday and Saturday daytime service sees trains operate every 12 minutes on the full length of both routes, giving a service every 6 minutes on the core section between South Gosforth and Pelaw. At peak times additional services operate on the Yellow route every 12 minutes between Monkseaton, South Gosforth and Pelaw and on the Green route every 12 minutes between Regent Centre and Pelaw, giving a service every 3 minutes in both directions on the core section. Evening services operate every 15 minutes after around 19.00. On Sundays trains start by running half-hourly and then from late morning operate every 15 minutes on both routes. The first trains on weekdays start running from 05.00 and finish at around 00.30.

The Metro system uses a fleet of 90 articulated twin Metro cars built by Metro-Cammell in Birmingham in the late 1970s to early 1980s, modelled on the German Stadtbahnwagen design. Of the three bogies, the outer two bogies are powered. Originally fitted with 84 seats, on mid-life refurbishment in the late 1990s/early 2000s the number of seats was reduced to 68 to allow for more standing space, with longitudinal instead of facing seating either side of the articulation. During a further major refurbishment between 2010 and 2015 at Wabtec Doncaster this was reduced to 64 seats to incorporate two dedicated wheelchair spaces. The crush loading capacity is quoted at 232 passengers per car.

There are half-width driving cabs at each end, with seats opposite them – giving passengers the unusual opportunity to sit "alongside the driver" and have a view of the line ahead. Because enough trains were originally built to allow 3-train operation – which was never implemented – there were enough trains in the fleet to cover both the Airport and South Hylton extensions.

Funding constraints meant that the fleet refurbishment only included 86 cars – 4001/02/40/83 were not refurbished and officially now form a reserve fleet, although do still see regular use. They can only operate with each other. 4022 has also been stored following an accident in the depot. It is currently at the Arriva Bristol Barton Hill depot where it initially went for assessment.

Standard fleet colours is a smart metallic silver, black and yellow livery, which can look very different depending on the lighting conditions. Original prototype cars 4001 and 4002 have also now received the standard livery, and the two other unrefurbished cars are in Advertising livery.

Metro cars always operate in pairs (in the early years cars operated singly on midweek evenings and winter Sundays). The Monday–Friday off-peak service requires 27 pairs (54 cars in traffic), 15 pairs on the Yellow route and 12 pairs on the Green route. An additional 12 pairs are used in the morning peak (nine Yellow/three Green) and 11 pairs in the evening peak (eight Yellow/ three Green), this means that from the fleet of 90 cars, 78 are required in the morning peak and 76 in the evening peak. Extra trains run as required at weekends and evenings in connection with football matches at both St James Park (Newcastle) and the Stadium of Light (Sunderland).

Metro also owns three battery electric locomotives and one Plasser & Theurer tamper. The locomotives are used for autumn railhead cleaning trains, depot shunting, engineering trains and to rescue failed Metro cars.

The system is publicly owned by Nexus (the Tyne & Wear Passenger Transport Executive) – Nexus also took direct control of the operation of Metro from DB Regio in 2017 (DB Regio having operated Metro since 2010). Nexus has recently started to consult on the future of the Metro. In June 2019 three bidders were shortlisted to build a new fleet of Tyne & Wear Metro cars – CAF, Hitachi and Stadler. The £500 million project to replace the fleet (including a 35-year maintenance contract) should see a new fleet of 42 4-car sets introduced on the network between 2022 and 2024 (the first is planned to be delivered in late 2021). The preferred bidder is due to be announced in January 2020. As part of the investment a second depot is to be built at Howdon, which will be used for overnight stabling whilst South Gosforth is rebuilt to house the new fleet of trains.

Ridership in 2018–19 was 36.4 million passengers, almost the same as the previous year. The Metro control centre and depot are both based at South Gosforth.

All Metro stations are unstaffed, with ticket machines. Some stations now have ticket gates. Single, return and all-day tickets are available. An all-day Metro ticket costs £5.20 at the time of writing. This ticket is also valid on the Shields ferry, linking North and South Shields. For observation purposes the busiest section of line is that between South Gosforth and Pelaw, which sees all trains. West Jesmond or South Gosforth are probably the best stations for an observation, and in around 1h45 all services can be seen here. Between West Jesmond and Jesmond an interconnecting line runs to Manors – this is retained for empty stock movements.

METRO-CAMMELL — 2-SECTION UNITS

Built: 1978–81 by Metropolitan Cammell, Birmingham (prototype cars 4001 and 4002 were built by Metropolitan Cammell in 1975 and rebuilt 1984–87 by Hunslet TPL, Leeds).
Wheel Arrangement: B-2-B.
Traction Motors: 2 x Siemens of 187 kW (250 hp) each.
Seats: 64 (* 68).
Weight: 39.0 tonnes.
Dimensions: 27.80 x 2.65 m.
Braking: Air/electro magnetic emergency track.
Couplers: BSI.
Maximum Speed: 50 mph.
Doors: Sliding plug.
Advertising Liveries: 4040 and 4083 – Emirates Airlines (red/various).

4001 *
4002 *
4003
4004
4005
4006
4007
4008
4009
4010
4011
4012
4013
4014
4015
4016
4017
4018
4019
4020
4021
4022 (S)
4023
4024
4025
4026
4027
4028
4029
4030
4031
4032
4033
4034
4035
4036
4037
4038
4039
4040 * **AL**
4041
4042
4043
4044
4045
4046
4047
4048
4049
4050
4051
4052
4053
4054
4055
4056
4057
4058
4059
4060
4061
4062
4063
4064
4065
4066
4067
4068
4069
4070
4071
4072
4073
4074
4075
4076
4077
4078
4079
4080
4081
4082
4083 * **AL**
4084
4085
4086
4087
4088
4089
4090

TYNE & WEAR METRO

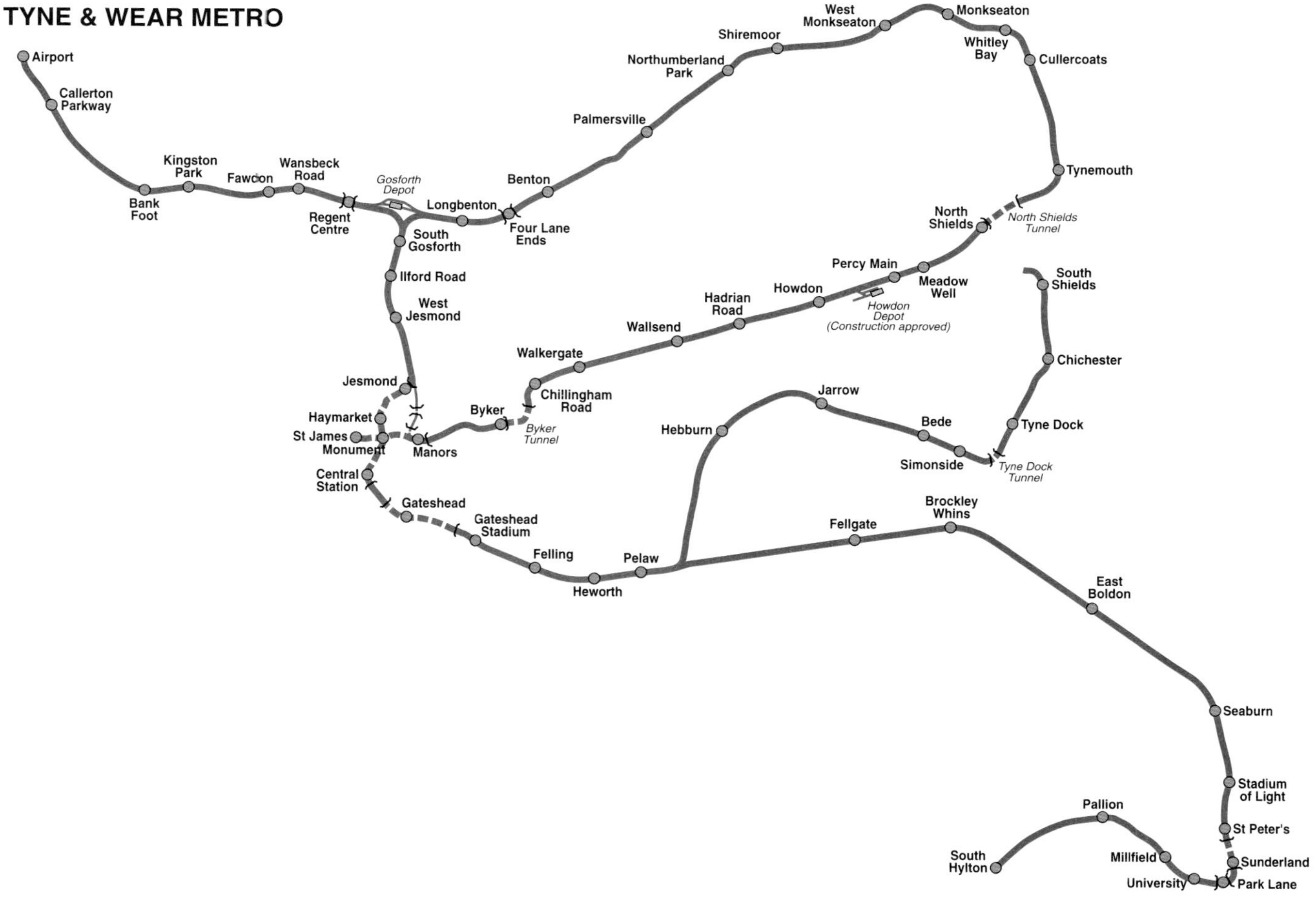

Airport
Callerton Parkway
Bank Foot
Kingston Park
Fawdon
Wansbeck Road
Regent Centre
Gosforth Depot
Longbenton
South Gosforth
Four Lane Ends
Benton
Palmersville
Northumberland Park
Shiremoor
West Monkseaton
Monkseaton
Whitley Bay
Cullercoats
Tynemouth
North Shields Tunnel
North Shields
Meadow Well
Percy Main
Howdon Depot (Construction approved)
Howdon
Hadrian Road
Wallsend
Walkergate
Chillingham Road
Byker Tunnel
Byker
Ilford Road
West Jesmond
Jesmond
Haymarket
St James
Monument
Manors
Central Station
Gateshead
Gateshead Stadium
Felling
Heworth
Pelaw
Hebburn
Jarrow
Bede
Simonside
Tyne Dock Tunnel
Tyne Dock
Chichester
South Shields
Fellgate
Brockley Whins
East Boldon
Seaburn
Stadium of Light
St Peter's
Sunderland
Park Lane
University
Millfield
Pallion
South Hylton

BATTERY ELECTRIC WORKS LOCOMOTIVES

Built: 1988–89 by Hunslet TPL, Leeds.
Battery:
Wheel Arrangement:
Traction Motors: 2 x Hunslet-Greenbat T9-4P of 67 kW (90 hp).
Dimensions: 6.30 m x 2.60 m.
Couplers: BSI.
Weight: 26.25 tonnes.
Maximum Speed: 30 mph.

Registered for operation on the national railway network as 97901–97903, and also now carry these numbers.

BL1 | BL2 | BL3

ON-TRACK MACHINE

MA60 Plasser & Theurer 08-275NX Switch & Crossing Tamper Built: 2013

▲ 4012+4072 depart from the fantastic station at Tynemouth with a service for St James on 4 April 2019. **Robert Pritchard**

▲ Last-built Metro car 4090 with 4008 arrive at Monkseaton with a service to St James via the Coast on 4 April 2019. **Robert Pritchard**

▼ Interior of refurbished Metro car 4060, looking towards the front. **Robert Pritchard**

▲ Battery electric works locomotive BL2 (97902) waits at Benton with an engineer's train from South Gosforth depot on 19 October 2018. BL1 was on the rear. **Kyle Allsopp**

▼ Plasser & Theurer 08-275NX Tamper MA60 is seen at Benton in the early hours of 13 January 2018 awaiting permission to enter a possession between Four Lane Ends and South Gosforth. **Kyle Allsopp**

3.10. WEST MIDLANDS METRO

Network: 13 miles. **Lines**: 1. **Depots**: 1 (Wednesbury). **First opened**: 1999.
www.westmidlandsmetro.com **System**: 750 V DC overhead.

Birmingham is the UK's second biggest city with a population of around 1.1 million. The Midland Metro tramway opened on 31 May 1999 with a 12-mile line between Birmingham's Snow Hill station and Wolverhampton, along the former Great Western Railway line via Handsworth, West Bromwich and Wednesbury (the GWR line terminated at Wolverhampton Low Level, with passenger services withdrawn in 1972 and freight in 1992). The old railway route has been rebuilt with new low floor platform stations. Approaching Wolverhampton Midland Metro leaves the old railway alignment, with around 1½ miles of street-running to the St Georges terminus, close to the main shopping area in the town (but not so convenient for the railway station).

A short but significant extension opened in 2015–16 when the tramway was extended through Birmingham city centre, first to Bull Street and then to a new terminus outside the recently redeveloped New Street station, called "Grand Central" after the adjacent shopping centre. This extension resulted in the original cramped Snow Hill terminus being abandoned as trams now use a new alignment to join street level near Bull Street.

From June 2018 Midland Metro has been branded "West Midlands Metro" and has been operated by Midland Metro Limited, now under public ownership as part of the West Midlands Combined Authority. Previously it has been privately operated by National Express, which also operates most of the bus services in the West Midlands.

The Monday–Saturday off-peak service frequency is every 8 minutes, increasing to every 6 minutes at peak times on weekdays. During the evening and on Sundays trams operate every 15 minutes. There are some short workings to and from Wednesbury Parkway for trams booked to run on or off the depot at peak times, but otherwise all trips operate the length of the route – the end-to-end journey time is 40 minutes across the 26 stops. 15 trams are required at peak times, with 12 trams needed to maintain the 8-minute off-peak service. Services operate from around 05.00 to midnight.

For the first 15–16 years of operation the tramway used 16 2-section Ansaldo T69 trams built in Italy. However, these were not popular and were unreliable, some spending lengthy periods out of traffic or being used for spares to keep other members of the class in traffic. As part of the upgrading of the tramway and the New Street extension, in 2012 20 new CAF trams were ordered, this was later increased to 21. The new CAF trams were delivered between October 2013 and July 2015, the first officially entering service in September 2014. The last of the T69s were withdrawn in August 2015. Most of the older trams were initially stored at the Long Marston site in Warwickshire, but in 2018 12 of these were sent for scrap. Three trams remain at Long Marston, whilst car 16 is still at Wednesbury depot for possible conversion to an engineering car.

The new vehicles are 5-section articulated bi-directional trams built by CAF in Spain. They provide the same number of seats as the older trams, but are longer, so have more room for standees. They are also 100% low floor. As well as the 52 seats there is quoted to be capacity for 156 standing passengers. At the same time as the new trams were delivered the depot at Wednesbury was extended to make it long enough to hold two of the longer trams.

The livery used for the original trams was blue and red, but a pink, silver and white scheme was introduced in 2007 (although only applied to four of the older trams). A revised version of this new livery was applied to the new CAF trams but was replaced in summer 2018 with a new two-tone blue livery as part of the rebranding of trams, trains and buses across the West Midlands. However, at the time of writing, this livery has only so far been applied in full to one tram.

Further extensions are planned and are being managed by the Midland Metro Alliance, a consortium that will "implement a 10-year programme of tram system enhancement works". The first extension will be a short link from the current terminus at Grand Central to Centenary Square via Victoria Square, with construction beginning in 2017 and opening planned for late 2019. Uniquely, for this section of line trams are planned to operate on battery power throughout, and all of the CAF trams are being modified to operate as such. The first car was returned to Spain for this modification, with the rest being completed at Wednesbury depot (all should be fitted by the end of 2019). Lithium ion batteries are being fitted to the trams' roofs and will be recharged by the overhead line along other parts of the route. The Midland Metro will be the

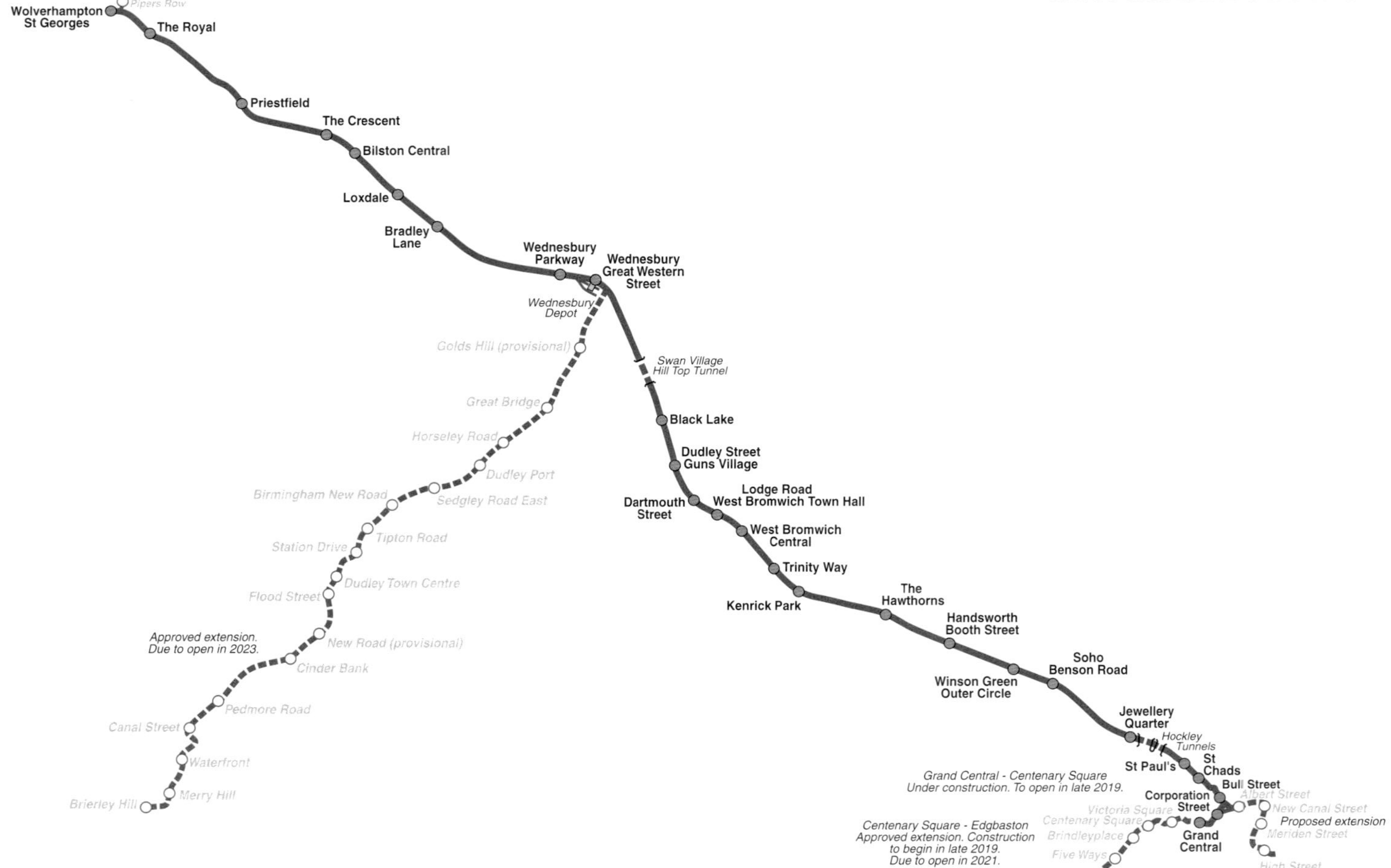

WEST MIDLANDS METRO
Under construction.
Due to open in 2020.
Wolverhampton
Railway Station
Pipers Row
Wolverhampton
St Georges
The Royal
Priestfield
The Crescent
Bilston Central
Loxdale
Bradley
Lane
Wednesbury
Parkway
Wednesbury
Great Western
Street
Wednesbury
Depot
Golds Hill (provisional)
Great Bridge
Horseley Road
Dudley Port
Sedgley Road East
Birmingham New Road
Tipton Road
Station Drive
Dudley Town Centre
Flood Street
New Road (provisional)
Cinder Bank
Approved extension.
Due to open in 2023.
Pedmore Road
Canal Street
Waterfront
Merry Hill
Brierley Hill
Swan Village
Hill Top Tunnel
Black Lake
Dudley Street
Guns Village
Dartmouth
Street
Lodge Road
West Bromwich Town Hall
West Bromwich
Central
Trinity Way
Kenrick Park
The
Hawthorns
Handsworth
Booth Street
Winson Green
Outer Circle
Soho
Benson Road
Jewellery
Quarter
Hockley
Tunnels
St Paul's
St
Chads
Bull Street
Corporation
Street
Grand
Central
Albert Street
New Canal Street
Proposed extension
Meriden Street
High Street
Deritend
Grand Central - Centenary Square
Under construction. To open in late 2019.
Victoria Square
Centenary Square
Brindleyplace
Five Ways
Centenary Square - Edgbaston
Approved extension. Construction
to begin in late 2019.
Due to open in 2021.

first commercial tram system in the UK to have catenary-free running, allowing trams to operate over short distances without the need for overhead wires, which it is said would have marred the architecturally sensitive area around Birmingham's historic town hall in Victoria Square.

At the other end of the route a short half-mile extension is under construction through Wolverhampton city centre to terminate at the railway station. This is planned to open in 2020, when it is planned that trams will alternately terminate at St Georges and Wolverhampton railway station. The next extension will run for around 1 mile from Centenary Square to Brindleyplace, Five Ways and Edgbaston, planned to open in late 2021. This will be followed (in 2023) by the 1-mile Birmingham Eastside extension. This will leave the Birmingham city centre route between Bull Street and Corporation Street and run to Digbeth High Street and High Street Deritend, including a stop at Curzon Street station (the planned new HS2 station for Birmingham). All of these routes will feature some catenary-free running.

Work is also now underway on opening a 7½ mile link south from Wednesbury to Dudley Port, Dudley and the Merry Hill shopping centre at Brierley Hill. This runs largely along the former South Staffordshire Railway alignment, with some street running in Dudley town centre. It is hoped to open this line by late 2023. Plans are also at an early stage for a 10-mile extension from Digbeth to Heartlands Hospital, the NEC/Birmingham Airport and the HS2 Interchange Station in north Solihull, with the hope that this will open in 2026.

The Centenary Square/Edgbaston and Wolverhampton extensions can be serviced by the existing fleet of 21 trams (this will mean increase the peak requirement to 19 trams). Further trams will be needed for the Eastside/Airport and Brierley Hill extensions and the West Midlands Combined Authority currently has a tender out for up to 50 new trams (a firm order for 18, with an option to take this up to 32), to be delivered between 2021 and 2026. Potentially this could give a fleet of 71 trains by 2026. Seven manufacturers have expressed an interest in the tender: Alstom, Bombardier, CAF, Chinese firm CRRC Qingdao Sifang, Turkish firm Durmazlar Makine Sanayi Ticaret, Škoda and Stadler. The four bidders with the highest scoring bid will make the Invitation to Negotiate shortlist, with contract award expected in September 2019. To operate the enlarged fleet additional depot and control room facilities will be required, along with new control and communications systems. The new fleet will also be equipped with batteries as-built.

Tickets range from just £1 for the popular city centre "short hop" single ticket (valid between Grand Central and Jewellery Quarter) to £4 for a tram only off-peak day ticket or £5.50 for a peak day ticket. An off-peak Metro+bus Daysaver ticket is £5.20, valid on National Express buses in the West Midlands and the Metro. Midland Metro is also included in the n-Network (West Midlands) Daytripper ticket (£6.70).

In terms of observing the trams, all services can be seen in Birmingham city centre, and the Wednesbury stops can also be interesting in terms of viewing trams running on and off the depot.

In 2018–19 Midland Metro carried 5.9 million passengers, up 2.5% on the previous year.

ANSALDO T69 — 3-SECTION TRAMS

Built: 1998–99 by Ansaldo Transporti, Italy.
Wheel Arrangement: Bo-2-Bo.
Seats: 52.
Dimensions: 24.00 x 2.65 m.
Couplers: Not equipped.
Doors: Sliding plug.
Traction Motors: 4 x 105 kW (140 hp).
Weight: 35.6 tonnes.
Braking: Regenerative, disc and magnetic track.
Maximum Speed: 43 mph.
Livery: Dark blue & light grey with a green stripe & red front end unless stated.
Other/Non-Standard Liveries: **MW:** Network West Midlands silver & pink.
0: Original Birmingham Corporation tram livery (cream & blue).

07, 10, 11 are stored at Long Marston. 07 and 10 are stored for use by UKTram for development work and 11 has been donated to Birmingham Museum.
16 is stored at Wednesbury depot for possible conversion to an engineering car.

07 (S) **MW**	10 (S) **MW**	11 (S) **0**	16 (S)

CAF URBOS 3 — 5-SECTION TRAMS

Built: 2013–14 by CAF, Zaragoza, Spain.
Wheel Arrangement: Bo-2-Bo.
Seats: 52.
Dimensions: 32.96 x 2.65 m.
Couplers: Albert.
Doors: Sliding plug.
Traction Motors: 8 x 65 kW (87 hp).
Weight: 41.0 tonnes.
Braking: Regenerative, disc and magnetic track.
Maximum Speed: 43 mph.
Standard livery: Network West Midlands silver & pink.
Other/Non-Standard Liveries: **B**: New West Midlands Metro two-tone blue.
0: Blue ends, partial adverts but otherwise the existing livery. 19 has vinyls to celebrate 20 years of the modern tramway.

* Fitted with roof-mounted lithium ion batteries to enable catenary-free operation.

No.		Livery
17	*	
18	*	0
19	*	0
20		
21	*	0
22		
23	*	
24		
25		
26		
27		
28	*	
29	*	
30		
31	*	B
32		
33		
34		
35	*	
36	*	0
37	*	

Names:

31 Cyrille Regis MBE 1958–2018
35 Angus Adams
37 OZZY OSBOURNE

▲ Stored Ansaldo T69 cars 07 and 11 at the Quinton Rail Technology Centre site at Long Marston on 20 June 2019. **Robert Pritchard**

▲ In a non-standard livery (original fleet livery with blue ends and partial advertising for Just Eat), 18 stands at the Grand Central terminus, outside Birmingham New Street station, with a service for Wolverhampton St Georges on 6 April 2019. The line will continue beyond the back of the tram to Centenary Square and eventually Edgbaston. **Robert Pritchard**

▼ In Wolverhampton, new West Midlands Metro blue-liveried 31 approaches The Royal with a service for Birmingham Grand Central on 6 April 2019. **Robert Pritchard**

▲ 35 is seen between Trinity Way and Kenrick Park with a service for Grand Central on 6 April 2019. The tower of the Holy Trinity church, West Bromwich can be seen on the skyline. **Robert Pritchard**

▼ Interior of Urbos 3 tram 17, showing the mix of facing and longitudinal seating. **Robert Pritchard**

APPENDIX I. LIST OF LOCATIONS

The following is a list of locations to accompany Sections 2: Preserved Underground Stock and Section 3.4.1: Glasgow Subway Preserved Stock. Ordnance Survey grid references are given for each location.

§ denotes site not generally open to the public.

	OS GRID REF
Alderney Railway, Mannez Quarry, Isle of Alderney, Channel Islands.	–
Bluebell Railway, Sheffield Park, near Uckfield, East Sussex.	TQ 403238
Bo'ness & Kinneil Railway, Bo'ness Station, Union Street, Bo'ness, Falkirk.	NT 003817
Buckinghamshire Railway Centre, Quainton Road Station, Aylesbury, Buckinghamshire.	SP 736189
Coopers Lane Primary School, Pragnell Road, Grove Park, Greater London.§	TQ 405728
Cravens Heritage Trains, Epping Signal Box, Epping Station, Epping, Essex.§	TL 462016
Epping Ongar Railway, Ongar Station, Station Road, Chipping Ongar, Essex.	TL 552035
Finmere Station, near Newton Purcell, Oxfordshire.§	SP 629312
Glasgow Museum Resources Centre, 200 Woodhead Road, Nitshill, Glasgow.§	NS 520601
Glasgow Riverside Museum, Pointhouse Quay, Yorkshill, Glasgow.	NS 557661
Hardingham Station, Low Street, Hardingham, Norfolk.§	TG 050055
Hope Farm (Southern Locomotives), Sellindge,near Ashford, Kent.§	TR 119388
Keighley & Worth Valley Railway, Haworth, near Keighley, West Yorkshire.	SE 034371
Kent & East Sussex Railway, Tenterden Town Station, Tenterden, Kent.	TQ 882336
London Transport Museum, Covent Garden, Greater London.	TQ 303809
London Transport Museum Depot, Gunnersby Lane, Acton, Greater London.	TQ 194799
London Underground, Acton Works, 130 Bollo Lane, Acton, Greater London.§	TQ 193798
London Underground, Hainault Depot, Thurlow Gardens, Ilford, Greater London.§	TQ 450920
London Underground, Northfields Depot, Northfield Avenue, Brentford, Gtr London.§	TQ 166789
London Underground, Ruislip Depot, West End Road, Ruislip, Greater London.§	TQ 094862
Longstowe Station, near Bourn, Cambridgeshire.§	TL 315546
Mangapps Railway Museum, Southminster Road, Burnham-on-Crouch, Essex.	TQ 944980
Men At Work, Units 90–95, Waterside Trading Centre, Trumpers Way, Hanwell, Greater London.§	TQ 153793
National Railway Museum, Shildon, Co Durham.	NZ 238256
Nene Valley Railway, Wansford Station, Peterborough, Cambridgeshire.	TL 093979
Tyne & Wear Fire Service Training Centre, Nissan Way, Barmston Meir, Washington, Tyne & Wear.§	NZ 329572
Village Underground, 54 Holwell Lane, off Bishopgate, Shoreditch, Greater London.	TQ 334823
Walthamstow Pumphouse Museum, The Pump House, Lowe Hall Lane, Walthamstow, Greater London.	TQ 362882
Woodlands, Elm Hill, Coppleridge, Motcombe, near Shaftesbury, Dorset.§	ST 846267

APPENDIX II. ABBREVIATIONS USED

ATO	Automatic Train Operation
BR	British Railways, later British Rail
C&SL	City & South London Railway
CCE&H	Charing Cross, Euston & Hampstead Railway
CLRC	Central London Railway Company
DLR	Docklands Light Railway
DR	District Railway
EMU	Electric Multiple Unit
GNP&BR	Great Northern, Piccadilly & Brompton Railway
LER	London Electric Railways
LNER	London & North Eastern Railway
LPTB	London Passenger Transport Board
LT	London Transport
LTE	London Transport Executive
LU/LUL	London Underground/London Underground Limited
MPU	Motive Power Unit
MR	Metropolitan Railway
MV	Milk Van
NET	Nottingham Express Transit
NTFL	New Tube for London
PMV	Parcels & Miscellaneous Van
PTE	Passenger Transport Executive
SPT	Strathclyde Partnership for Transport
SSL	Sub-Surface Line
SYPTE	South Yorkshire Passenger Transport Executive
SYSL	South Yorkshire Supertram Limited
TfGM	Transport for Greater Manchester
TfL	Transport for London
tph	Trains per Hour
TRV	Track Recording Vehicle
W&C	Waterloo & City Line
W&CR	Waterloo & City Railway
WMCA	West Midlands Combined Authority

LIVERY CODES

AL	Advertising Livery
0	Other (non-standard) Livery

VEHICLE TYPES

B	Brake
D	Driving
F	First
M	Motor
O	Open
S	Second/Standard
T	Trailer (or Third for hauled stock) (TT=Trailer Third; DTT=Driving Trailer Third)

APPENDIX III. WEIGHTS & MEASUREMENTS

AC	Alternating current
Ah	Amp-hour
cwt	Hundredweight
DC	Direct current
hp	Horsepower
kN	Kilonewtons
kW	Kilowatts
lbf	Pounds force
lbf/sq. in.	Pounds force per square inch

m	Metres
mm	Millimetres
mph	Miles per hour
tonnes	Metric tonnes
tons	Imperial tons
TOPS	Total Operations Processing System
V	Volts

APPENDIX IV. PRIVATE MANUFACTURERS

ABB	ASEA Brown Boveri, (later ABB, now Bombardier Transportation)
Adtranz	ABB/Daimler-Benz Transport
Alstom	Alstom Transportation
Ansaldo	AnsaldoBreda (now Hitachi Rail Italy)
AP	Aveling & Porter
Ashbury	Ashbury Carriage & Iron Company
BM	Brown Marshall
BN	BN Construction (now Bombardier Transportation)
Bombardier	Bombardier Transportation
BP	Beyer Peacock and Company
BR	British Railways (later British Rail)
BRCW	Birmingham Railway Carriage & Wagon Company
BREL	British Rail Engineering Limited (later BREL, then ABB, now Bombardier Transportation)
Brush	Brush Traction
BTH	British Thomson-Houston Company
CAF	Construcciones y Auxiliar de Ferrocarriles
CL	Cammel Laird
Clayton	Clayton Equipment
Cowans	Cowans, Sheldon and Company
Cravens	Cravens Railway Carriage & Wagon Company
EE	English Electric Company
Firema	Firema Trasporti
GEC	General Electric Company (later GEC, then GEC-Alsthom, now Alstom)
GRCW	Gloucester Railway Carriage & Wagon Company
HB	Hunslet-Barclay Limited (now Wabtec Corporation)
Hunslet	Hunslet Engine Company (now Wabtec Corporation)
Hurst	Hurst Nelson & Company
Kawasaki	Kawasaki Rail Car, Inc
Mather	Mather & Platt
Matisa	Matisa Matériel Industriel SA
MC	Metropolitan-Cammell Carriage & Wagon Company (later Metro-Cammell)
MV	Metropolitan-Vickers
Neasden	Metropolitan Railway Neasden Works
Oldbury	Oldbury Railway Carriage & Wagon Company
Plasser	Plasser & Theurer
RFS	RFS Industries (now Wabtec Corporation)
RR	Rolls-Royce
Schoma	Christoph Schöttler Maschinenfabrik GmbH (now Schöma)
Siemens	Siemens AG
Stadler	Stadler Rail
VL	Vossloh Locomotives
Vossloh	Vossloh Rail Vehicles
Wickham	D Wickham and Company